LATIN AMERICAN POLITICAL, ECONOMIC, AND SECURITY ISSUES

COLOMBIA

SOCIAL, ECONOMIC AND ENVIRONMENTAL ISSUES

Latin American Political, Economic, and Security Issues

Additional books in this series can be found on Nova's website under the Series tab.

Additional E-books in this series can be found on Nova's website under the E-book tab.

Global Economic Studies

Additional books in this series can be found on Nova's website under the Series tab.

Additional E-books in this series can be found on Nova's website under the E-book tab.

LATIN AMERICAN POLITICAL, ECONOMIC, AND SECURITY ISSUES

COLOMBIA

SOCIAL, ECONOMIC AND ENVIRONMENTAL ISSUES

GREGORY LAVIGNE
AND
CATHERINE COTE
EDITORS

NOVINKA

New York

For permission to use material from this book please contact us:
Telephone 631-231-7269; Fax 631-231-8175
Web Site: http://www.novapublishers.com

Library of Congress Cataloging-in-Publication Data

Colombia : social, economic and environmental issues / editors: Gregory Lavigne and Catherine Cote.
pages cm
Includes index.
ISBN 978-1-62417-848-1 (soft cover)
1. Colombia--Environmental conditions. 2. Environmental policy--Colombia. 3. People with disabilities--Colombia--Social conditions. 4. People with disabilities--Government poicy--Colombia. I. Lavigne, Gregory. II. Cote, Catherine.
GE160.C7C65 2013
304.209861--dc23
2012051468

Published by Nova Science Publishers, Inc. † New York

CONTENTS

PREFACE

In this book, the authors present current research in the study of the social, economic and environmental issues facing Colombia today. Topics discussed include the major environmental issues confronting Bogota's wetlands by analysis of their biodiversity, physicochemistry and eutrophication; elasmobranch biology and diversity in the marine and continental environments of Colombia; social and technological advances for people with disabilities in Colombia; and the climate and natural hazards of the Colombian Caribbean region.

Chapter 1 - Wetlands have been regarded as hotspots of biodiversity and important sites for nutrient biogeocycling. Localized on the eastern cordillera of the Andes, the savanna of Bogotá, Colombia was once a great complex of wetlands. The present-day wetlands are the remnants of a large wetland area that developed around 27,000 years ago when the Bogotá basin became over-filled with sediments. After centuries of human intervention by urban establishment and growth, the original wetland area has been severely reduced from 50,000 ha in 1950 to the current 800 ha. Presently, the 11 to 13 fragmented wetland areas are scattered throughout the city. They include Santa María de Lago, Córdoba, Juan Amarillo, Capellanía, La Vaca, el Burro, Techo, Tibanica, Jaboque, Guaymaral, Torca, La Conejera and Meandro del Say. Most of them are home to many species of terrestrial and aquatic vertebrates, invertebrates, plants and other organisms. Many of these species are endemic or migratory, some of which are threatened, or endangered. Most of the above wetlands have been severely degraded and their original areas have been reduced mainly by dumping of solid waste and sewage and by illegal housing. Other environmental issues include water pollution, sound pollution, invasive species introduction and habitat loss. Governmental actions have been addressed. The city of Bogotá has produced laws and decrees to preserve the wetlands; however, policy has not been taken seriously and corruption has limited the efforts of environmentalists and wetland defenders. There are a number of non-governmental organizations, scientists and interested citizens who have been working to preserve these unique places. There remains the possibility to preserve these wetlands from destructive anthropogenic intervention. Bogotá's wetlands might be able to cope and coexist with today's urban development if actions and the appropriate conscious attitude is exercised by citizens, the public and private sectors.

This chapter illustrates the major environmental issues confronting Bogotá's wetlands by analysis of their biodiversity, physicochemistry and eutrophication. It discusses the environmental and policy issues surrounding water pollution, sound pollution, and landfilling, as it addresses environmental laws and restoration action plans that have been carried out to preserve the city's wetlands. The authors encourage the general public, young researchers and senior scientists to promote actions for preservation of these magnificent ecosystems.

Chapter 2 - There is increasing interest in elasmobranch biology and diversity in the marine and continental environments of Colombia. The number of elasmobranch species are estimated to be about 124, but this number is likely to increase as more elasmobranchs are reported. Of these, eight species are freshwater stingrays. Populations of sharks and rays have been reducing in Colombia since the 1960s. Shark, ray and chimera fishery landings were about 157 T from 1958 to 1970 which reduced to about 76 T between 2001 and 2009 in the Colombian Atlantic Ocean. Shark, ray and chimera catches have diminished from 539 T from 1981 to 1990 to 353 T between 2001 and 2009 in the Colombian Pacific Ocean. However, catches from illegal fishing vessels of national and international origin are unreported so the actual rate of attrition is unknown. Broad taxonomic information has been gathered by the Colombian environmental and governmental institutions since 1957. Little research has been conducted on elasmobranch feeding ecology in the Colombian Atlantic and Pacific Oceans when compared to the total amount of elasmonbranch species reported for Colombia. Most studies in marine environments have focused on demersal elasmobranchs. Elasmobranch reproduction has been reported off the Atlantic and Pacific coasts and on the Magdalena River. Shark movements have been studied off the Colombian Pacific Ocean. Evidence exists from both scientists and fishermen that sharks venture into Colombian rivers. Colombia recently developed a National Action Plan for Conservation and Management of sharks, rays and chimeras. Elasmobranch fisheries in Colombia are not well documented. According to

FAO (Food and Agriculture Organization) fisheries statistics, both Pacific and Atlantic oceans show a decrease in elasmobranch landings since the 1960s. More research is needed by Colombian scientists to better understand elasmobranch biology in the waters of northern South America.

Chapter 3 - Having as a reference the work developed by the Center of Stimulation, Leveling, and Development CEDESNID for the habilitation/rehabilitation and social inclusion of the people with the condition of cognitive disability, this chapter has the purpose of doing a revision of the current status of cognitive disability in Colombia, to afterwards present the impact that can have the technological innovation in this field. For this it will be done a recount of the principal tendencies and lines of action at international level about the approach of the disability as a generality and of the cognitive/intellectual disability as a particular category. Afterwards, it will be made a revision of the most outstanding tendencies in Colombia. Finally, it will be presented the PreLingua tool as an example of the technological innovation of high social impact to be applied in the elimination of the involved barriers in the cognitive disability and as seed project for the innovation in Colombia.

Chapter 4 - This chapter describes the climate of the northern Colombia called the Colombian Caribbean region. The climate in this area of Colombia is driven by the movement of the inter-tropical convergence zone (ITCZ), which produces two seasons: a dry (low rainfall) season from November to May and a rainy season (high rainfall) from April to October, with a season of transition called 'veranillo de San Juan' between June and July. Also, the vulnerability to natural hazards is described; with emphasis in phenomena such as hurricanes and storms, tornadoes and cold fronts. Although the Colombian Caribbean is not known as a natural disaster area, hurricanes (e.g., Joan in 1988 and Lenny in 1999), tornadoes, windstorms and cold fronts, can affect the climate of this Caribbean area, causing damage to infrastructure and loss of life in some cases.

In: Colombia
Editors: G. Lavigne and C. Cote
ISBN: 978-1-62417-848-1

Chapter 1

BOGOTÁ'S URBAN WETLANDS: ENVIRONMENTAL ISSUES

Fernando Ramírez[1,*], Thomas Lee Davenport[2,#] and Jose Kallarackal[3,¥]

[1]Facultad de Ciencias, Pontificia Universidad Javeriana, Bogotá, Colombia
[2]Director of Research and Development, VivaFresh Technologies, Florida City, FL, United States of America
[3]Sustainable Forest Management Division, Kerala Forest Research Institute, Peechi, Thrissur, Kerala, India

ABSTRACT

Wetlands have been regarded as hotspots of biodiversity and important sites for nutrient biogeocycling. Localized on the eastern cordillera of the Andes, the savanna of Bogotá, Colombia was once a great complex of wetlands. The present-day wetlands are the remnants of a large wetland area that developed around 27,000 years ago when the Bogotá basin became over-filled with sediments. After centuries of human intervention by urban establishment and growth, the original wetland area has been severely reduced from 50,000 ha in 1950 to the current 800 ha. Presently, the 11 to 13 fragmented wetland areas are scattered throughout the city. They include Santa María de Lago,

[*] E-mail: fernando.ramirez@javeriana.edu.co., fermp44@yahoo.com.
[#] E-mail: tldav@ufl.edu.
[¥] E-mail: jkallara@gmail.com.

Córdoba, Juan Amarillo, Capellanía, La Vaca, el Burro, Techo, Tibanica, Jaboque, Guaymaral, Torca, La Conejera and Meandro del Say. Most of them are home to many species of terrestrial and aquatic vertebrates, invertebrates, plants and other organisms. Many of these species are endemic or migratory, some of which are threatened, or endangered. Most of the above wetlands have been severely degraded and their original areas have been reduced mainly by dumping of solid waste and sewage and by illegal housing. Other environmental issues include water pollution, sound pollution, invasive species introduction and habitat loss. Governmental actions have been addressed. The city of Bogotá has produced laws and decrees to preserve the wetlands; however, policy has not been taken seriously and corruption has limited the efforts of environmentalists and wetland defenders. There are a number of non-governmental organizations, scientists and interested citizens who have been working to preserve these unique places. There remains the possibility to preserve these wetlands from destructive anthropogenic intervention. Bogotá's wetlands might be able to cope and coexist with today's urban development if actions and the appropriate conscious attitude is exercised by citizens, the public and private sectors.

This chapter illustrates the major environmental issues confronting Bogotá's wetlands by analysis of their biodiversity, physicochemistry and eutrophication. It discusses the environmental and policy issues surrounding water pollution, sound pollution, and landfilling, as it addresses environmental laws and restoration action plans that have been carried out to preserve the city's wetlands. The authors encourage the general public, young researchers and senior scientists to promote actions for preservation of these magnificent ecosystems.

INTRODUCTION

Wetlands are areas that are inundated or saturated by surface or ground water at a frequency and duration sufficient to support, and that under normal circumstances do support, a prevalence of vegetation typically adapted for life in saturated soil conditions (taken from US Environmental Protection Agency Regulations). Wetlands generally include swamps, marshes, bogs, and similar areas. For years, wetlands were viewed as unproductive wastelands in many parts of the world. Their natural values were measured against the value of their use as dry land. Because the natural functions of wetlands were not well understood, the accepted practice was to drain them or fill them with soil in order to make the land more "useful" to humans. The human view of the wetlands is, on occasion, focused too narrowly on the economic value of the area. Wetlands, however, have functions that are overlooked as a vital part of the ecosystem. Wetlands are now recognized as important features in the landscape that provide numerous ecosystem services for people and for fish and wildlife. Some of these services, or functions, include protecting and improving water quality, providing fish and wildlife habitats, storing floodwaters, and maintaining surface water flow during dry periods. They are also important in enhancing the aesthetics of landscapes and are significant areas of biological productivity. These ecosystem services, considered valuable to societies worldwide, are the result of the inherent and unique natural characteristics of wetlands. However, the impacts, whether natural or anthropogenic can reduce the value of wetlands. Wetlands function like natural sponges, storing water and slowly releasing it. This process slows the water's momentum and erosive potential, reduces flood damages, and allows for ground water recharge, which contributes to base flow to surface water systems during dry periods.

After being slowed by a wetland, water moves around plants, allowing the suspended sediment to drop out and settle to the wetland floor. Nutrients from fertilizer application, manure, leaking septic tanks, and municipal sewage that are dissolved in the water are often absorbed by plant roots and microorganisms in the soil. Other pollutants stick to soil particles. In many cases, this filtration process removes much of the water's nutrient and pollutant load by the time it leaves a wetland. As already mentioned, wetlands are some of the most biologically productive natural ecosystems in the world, comparable to tropical rain forests and coral reefs in their productivity and the diversity of species they support. Abundant vegetation and shallow water provide diverse habitats for fish and wildlife. Aquatic plant life flourishes in the nutrient-rich environment, and energy converted by the plants is passed up the food chain to fish, waterfowl, and other wildlife and to us as well.

In recent years, wetlands have received a large amount of public attention. As a result, increased scientific study of these unique ecosystems at the global, national, state, and local levels has taken place. These studies have revealed the important biological, hydrological, terrestrial, and ecological functions wetlands perform. Knowledge of these functions has improved the recognition of wetlands as important, valuable ecosystems.

In this chapter, we have examined 12 wetlands of Bogotá, the largest city of Colombia. They include Santa María de Lago, Córdoba, Juan Amarillo, Capellanía, La Vaca, el Burro, Techo, Tibanica, Jaboque, Guaymaral-Torca, La Conejera and Meandro del Say. Taken individually, they may appear to be small, but form a strong network of wetlands when taken collectively. The present-day wetlands are the remnants of a large wetland area that developed around 27,000 years ago when the Bogotá basin became over-filled with sediments.

Early indigenous people and Spaniard conquerors were the first human groups to interact with the magnificent wetlands of Bogotá's savannah. After centuries of human intervention by urban establishment and growth, the original wetland area has been severely reduced from 50,000 ha in 1950 to the current 800 ha. Most major wetlands are home to many species of terrestrial and aquatic vertebrates, invertebrates, plants and other organisms. Many of these species are endemic or migratory, some of which are vulnerable, or endangered.

Most of these wetlands have been severely degraded and their original areas have been reduced mainly by dumping solid waste and sewage and by illegal landfills and housing. Other environmental issues include water pollution, sound pollution, invasive species introduction and habitat loss.

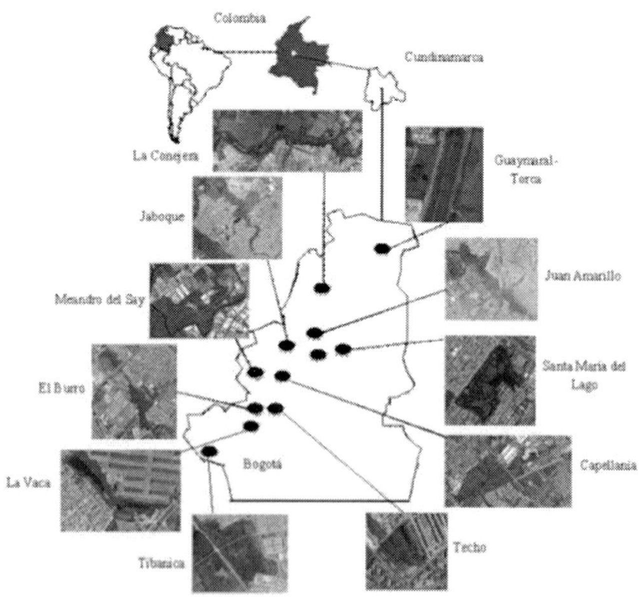

Photos from Google Earth 2012.

Figure 1. Wetlands within Bogotá city.

Wetland fragmentation, sewage inputs, garbage and solid residues disposal, area loss, hydric deficit, predator invasive species, adjacent urban impacts, highway impacts, lack of control and wetland security, lack of community interest, etc. (SDA, 2010c) have made Bogota's wetlands a highly vulnerable ecosystem.

Governmental actions have been addressed in this chapter. The city of Bogotá has produced laws and decrees to preserve the wetlands; however, policy has not been taken seriously and corruption has limited the efforts of environmentalists and wetland defenders. The plan establishes environmental zones. These include protection zones, conservation zones, restoration zones and control zones. The plan has five areas: 1) applied research, 2) education, participation and communication, 3) restoration, 4) management and sustainable use and 5) Inter-institutional management (SDA, 2010c).

There are a number of non-governmental organizations, scientists and interested citizens who have been working to preserve these unique places. There remains the possibility to preserve these wetlands from destructive human intervention. Bogotá's wetlands might be able to cope and coexist with today's urban development if actions and the appropriate conscious attitude is exercised by citizens, the public and private sectors. There remains a hope and an illusion that these wetlands and their marvellous inhabitants will not be doomed and the efforts of the many institutions and people interested in their conservation will be successful.

This chapter illustrates the major environmental issues confronting Bogotá's wetlands by analysis of their biodiversity, physicochemistry and eutrophication. It discusses the environmental and policy issues surrounding water pollution, sound pollution, and landfilling, as it addresses environmental laws and restoration action plans that have been carried out to preserve the city's wetlands. The authors encourage the general public, young researchers and senior scientists to promote actions for preservation of these magnificent wetlands.

THE WETLANDS WITHIN THE CITY

Located at the western Cordillera of Colombia, the savanna of Bogotá extends over 427,000 Ha and at an elevation of 2,600 m.o.s.l. This vast savanna was covered by large swamps and weedy lakes during the Pleistocene (Van Geel and Van der Hammen, 1973).

Source: Google Earth 2012.

Figure 2. The Santa María del Lago Wetland.

Indigenous cultures made channels and levees and managed the water flow within the wetlands (Van der Hammen *et al.*, 2008). By the time of the Spanish conquest in the early 1500, the savanna had the character of a wetland (Fjeldså, 1984). By 1538, the primitive sewage draining system of Bogota flowed directly into the Bogotá River and the surrounding lagoons and wetlands (Van der Hammen *et al.*, 2008). Land-use expansion increased in the 19th century, and the wetlands lost much of their original area. Roads, highways and urbanization projects constructed in the 1930-1950s reduced and fragmented Bogota´s wetlands (Van der Hammen *et al.*, 2008). This is the case of the large Tintal lake that was fragmented into the Tibanica, La Vaca, El Burro, and Techo wetlands. Massive landfilling, invasion and construction within the wetlandsbegan by the 1960s (Van der Hammen *et al.*, 2008). Today, Bogotá is home to more than 200 water systems; among them wetlands, streams, canals and rivers (Cerón and Ramírez, 2011). Today, the total wetland area within the city is about 500 Ha (Pinilla, 2010) and includes 12 different wetlands (Figure 1). Wetlands originally covered 50,000 Ha and have been reduced to 800 Ha (DAMA, 2006).

Figure 3. Santa María del Lago facilities. A) Fenced entrance. B) Fenced perimeter. C) Sidewalks for pedestrians. This side-fence borders the wetland's perimeter. D) Entrance guard and signs. E) Guidelines sign. F) Bird identification chart.

This reduction has been attributed to human activity, namely land filling, a process that has increased urbanization. Urbanization has been carried out extensively throughout Bogota's savanna. Much of the wetland area loss has been associated with the disappearance of open water and replaced by dominant vegetation cover (Van der Hammen *et al.*, 2008).

Eutrophication is one of the major impacts that Bogotás wetlands face. Overall, Bogota´s wetland ecosystems are enriched with ions, nutrients, and organic matter in greater or lesser degrees (Pinilla, 2010). Urban activity has exacerbated these conditions. Information on the physicochemical characteristics of Bogotá's wetlands is scarce, making comparisons with earlier data difficult (Pinilla, 2010).

Photos by Fernando Ramírez September 2012.

Figure 4. Santa María del Lago' s aquatic plants. A) Water-terrestrial interface. Note *Schoenoplectus* californicus. B) Large community of *Thypha latifolia* (tall) and *Bidens laevis* (smaller). C) Closer look. D*) Bidens laevis*. E) The water fern *Azolla filiculoides*. F-G) *Polygonum hydropiperoides*. H) *Cyperus rufus*.

Photos by Fernando Ramírez. September 2012.

Figure 5. Santa María del Lago Birds. A) The diver *Podilymbus podiceps* swimming in open waters. B) *Fulica americana*. C) closer view. D) front view. E) grooming. F) nibbling among aquatic plants.

Wetlands with the greatest deterioration are mainly supported by taxa that are more resistant to pollution, whereas the best-preserved wetlands have greater species diversity (Pinilla, 2010). Other environmental problems that wetlands face include invasive species such as dogs that prey on birds and other organisms. People that access the wetlands are known to cause disturbance or habitat destruction (Rosselli and Stiles, 2012). Solid waste disposal and landfilling are other important factors that have contributed to the disappearance of Bogotá's wetlands. They can be dangerous places for people to visit. Thieves who use these areas for sanctuary frequently rob those that venture into the wetlands. Naturalists, and visitors therefore, are recommended to be escorted by local police patrol. Police consider the wetlands to be places of great danger to people because of the lack of entry control (access is granted to any person), most are not patrolled and are often classified as ' forgotten' . Unfortunately, the government has paid little attention to the city's wetlands and, indeed they remain largely forgotten.

Bogotá's wetlands are greatly influenced by the climatic dynamics of the savanna. The climate in Bogotá has a bimodal precipitation regime with heavier rains occurring in March and April as well as in October and November (Vasconcellos *et al.*, 2011). Temperatures range from 12 °C to 16 °C (average 12.9 °C) with wind speedvarying between 2.0 and 13.0 km h^{-1} (DAMA, 2006). However, variable climatic conditions due to climate change have changed precipitation patterns making it difficult to assign distinct dry and rainy seasons as these conditions now occur randomly during the year.

Despite the fact that most South American countries have signed the Ramsar convention (an intergovernmental treaty that provides the framework for national action and international cooperation for the conservation and wise use of wetlands and their resources), efforts to generate basic data have been insufficient. This lack of information has hindered the formulation of wetland-friendly policies allowing the sustainable management of these areas (Junk, 2012). Wetland conservation actions began in 1992. Since then, Colombia has developed laws and regulations for wetland protection and conservation (Table 1). Colombia's national environmental system was created after a law passed in 1993 (Law 99 of 1993). This environmental system, as part of the government, has been in charge of developing a conservation and sustainable strategy for wetlands (Van Der Hammen *et al.*, 2008).

Table 1. Environmental wetland laws proposed by the Colombian government

Regulation	Law (s)	Date
Adhesion to Ramsar Convention	357	1997
Biological Diversity Agreement	165	1994
Water public domain	2811	1974
Environmental protection	1421	1993
Wetlands as ecological parks in Bogotá	99/93	1993

Source: Google Earth 2012.

Figure 6. The Cordoba Wetland. Note Suba Highway dividing the wetland (from top to bottom toward the middle of the picture).

Colombia becamepart of the Ramsarconvention in 1997 (Law 357, 21 January, 1997) (Van Der Hammen *et al.*, 2008). Currently, wetlands hold the status of areas of ecological importance within the public domain. According to Colombia's constitution (Law 79), therefore, the government has the obligation to protect and conserve ecologically important areas. Despite the mandate and the political status of wetlands, few actions have been effectuated and governmental institutions are not as involved as they should be. Much more control and management is required for a measurable improvement to take place. The Eastern Andes wetland complex of Colombia, isolated from other similar highland wetlands in the Andes represents a major center of evolution of birds and other organisms such as fish and plants (Fjeldsa, 1985; Schmidt-Mumm, 1998; Calvachi, 2003; Rosselli and Stiles, 2012). Aquatic plant species from the wetlands account for about 68 species (Guzmán, 2012). Aquatic birds comprise 72 species (Van der Hammen *et al.*, 2008). Some bird species are now extinct for example the Colombian Grebe (*Podiceps andinus*) and two subspecies Niceforo' s Pintail (*Anas georgica niceforoi*) and *Polystictus pectoralis bogotensis* (Rosselli and Stiles, 2012). Other endemic species (*Rallus semiplumbeus and Cistothorus apolinari*) and two subspecies (*Ixobrychus exilis bogotensis* and *Gallinula melanops bogotensis*) are endangered (Renjifo *et al.*, 2002). Within Bogota's wetlands, water-bird richness is directly related to wetland area (Rosselli and Stiles, 2012).

Figure 7. Governmental signs. A) Restoration sign. B) Species sign.

Photos by Fernando Ramírez from bridge over channel.

Figure 8. Channel that enters the Cordoba wetland. A) channel that passes at the lower Suba highway (Avenida suba). Note *Sambucus peruviana* trees on each side. B) Further upstream. C) channel at the upper Suba highway. D) detail. Note that the black color is due to high pollutant content and noxious odor.

Likewise, wetlands with large proportions of emergent vegetation tend to support more water bird species of foraging guilds (Rosselli and Stiles, 2012).

Photos by Fernando Ramírez.

Figure 9. The Cordoba wetland aquatic plants. A) *Thypha latifolia* (tall) and *Bidens laevis* (small) near the water's edge. B) Aquatic plant community. C) Closer look. D) Dwarf water plants. E) *Hydrocotyle ranunculoides* to the right. F) Plant community detail: photos taken from the tenth floor of a nearby building.

None of the wetlands of Bogotá have been thoroughly studied. Most investigations have been conducted by governmental institutions such as Bogotás city office (SDA), Cunadinamarca state office (Corporacion Autónoma Regional de Cundinamarca, CAR) and Bogota's Water Company (Empresa de Acueducto y Alcantarillado de Bogotá, EAAB) as well as other private organizations and NGOs.

Photos by Fernando Ramírez.

Figure 10. Severe water loss in the lower part of the wetland. A) Water loss from an area that was previously flooded as a consequence of water deficit within the wetland below Suba highway. B) Close up view. Note that some water plants still remain. C) another "dry" area. D) Aquatic plants that start to wilt as a consequence of water deficit.

Most of the information compiled by these organizations is largely preliminary and does not provide a consistent view of the current wetland condition. Both aquatic and terrestrial communities need more research. Recognizing the need for more studies in scientific fields, such as limnology, ecology, connectivity, restoration, conservation, and physiology to better understand these wonderfully unique places, this chapter was written in appreciation of the wetlands' complexity.

Santa Maria del Lago Wetland

General Information

Santa María del Lago is located to the northwest of Bogotá. This wetland is found a few blocks from Boyacá Highway and 80th street, both considered major highways that connect the north-south and north-west areas of the city respectively. It is nested upon minor roads and blocks of houses and buildings that are part of a city zone referred to as Engativá. Santa Maria del Lago has been estimated to occupy approximately 4 Ha (Figure 2) (Moreno *et al.*, 2005; Pinilla, 2010) to10.4 Ha (DAMA, 2006). It was recently determined, however, by satellite imagery that Santa María del Lago's total wetland area is 5.42 Ha. (Rosselli and Stiles, 2012). The wetland´s perimeter has been fenced by Bogota's city office in order to control and protect it from possible land-filling operations, and to provide visitors a safe environment (Figure 3A and B). Bogota's city office and Water Company have built an inner pavement track that surrounds the wetland´s perimeter (Figure 3C). This track is used by naturalists, bird watchers, and other visitors that often use it for jogging, running or related sports. Within the wetland, there are a series of signs informing visitors of prohibited activities such as swimming, fishing, and garbage disposal (Figures 3D and E). Billboards point out the most important bird species found within species (Figure 3F). Trees around the wetland have been identified with signs below them in an effort to inform visitors. Local guards patrol the wetland. This has been the only wetland to have a good patrol system.

Photo by Fernando Ramírez.

Figure 11. Garbage disposed about five blocks from Suba highway near the water´s edge.

Source: Google Earth 2012.

Figure 12. La Conejera Wetland.

The local community, Bogota's city office and the Water Company conduct regular maintenance on the wetland. These measures include aquatic-plant removal and landscape management. Some of these measures are controversial because of modification or disruption of local plant distributions, for example, aquatic plant removal has caused impacts on growth dynamics, and currently there is insufficient knowledge of the wetland's environmental issues.

Santa Maria de Lago is considered a type D hydric flow wetland according to the hydric system classification system proposed by Van der Hammen *et al.* (2008) but is characterized as a wetland without hydric flow due to lack of water channels flowing in or out of the system. Thus, wetland is mainly fed by underground springs. This spring-fed system is located on a basin that often reminds the common peasant of an oasis within the city. Due to the spring-origin of the wetland, water levels rarely experience low levels. Most of the wetland has an average of 2 – 3 meters in depth.

Biodiversity

The wetland's perimeter can be divided into a terrestrial, aquatic, and an intermediate zone between the terrestrial and aquatic zones. The limits imposed within these zones are clearly visible (Figures 4A-H). Different zones have been classified between the terrestrial and aquatic environments for more information see Van der Hammen *et al.* (2008) and Rosselli and Stiles (2012). The terrestrial zone on the wetland's water-edge margin has been planted with native and introduced tree species. Native trees include locally known Aliso

(*Alnus acuminata*), Elder (*Sambucus niger*), and Willow (*Salix humbodtiana*), etc. There are also a number of introduced tree species such as *Eucalyptus*, *Acacia* and *Fraxinus*, etc. Introduced vegetation is represented by the grass, *Pennisetum clandestinum*, and herbaceous plants such as dandelion (*Taraxacum* sp.) and the American black nightshade *Solanum americanum* among many others.

The aquatic zone is characterized by water-plant communities that are well established toward the margin and in some cases in the open water. These communities are represented by 17 species (Figures 4A-H) (Pinilla, 2010). Aquatic plants are recognized by their capacity to absorb nitrogen, phosphorous and organic compounds, as well as heavy metals from the water. Aquatic plant abundance has been quantified using frequency of occurrence (FO) in Santa Maria del Lago. This is defined as a proportion; the abundance of any given species in respect to the abundance of all species. According to Pinilla (2010), the most abundant aquatic plant is smooth beggarticks (*Bidens laevis*) (FO= 57.25). This plant is found flowering along the margins of Santa María and forms dense Matts that are easily identified by the yellow-colored flowers (Figures 4B-D). The broadleaf cattail (*Typha latifolia*) (FO = 50), found along the margin´s edge, also grows densely in portions of the wetland forming small ' islands' that are the habitat of many invertebrates and birds that frequent the wetland (Figures 4B-C). The common duckweed (*Lemna minor*) (FO = 47), least duckweed (*Lemna minuta*) (FO = 44.93), and duckweed (*Spirodela intermedia*) (FO = 44.37), are small floating plants that can form a carpet-like cover on the water's surface. These plants are manually removed (when spreading toward the water free zone) from Santa Maria, because they tend to cover the surface and have been claimed to cause light reduction and oxygen depletion because underlying water layers cannot mix with upper water strata. These duckweeds provide shelter and protection for a number of aquatic invertebrates such as water fleas (*Daphnia*), aquatic beetles (Coleoptera) and larval stages (Ramírez personal observation). The water fern (*Azolla filiculoides*) (FO = 36.62) is a floating plant distributing similarly to the duckweeds as a carpet on the water's surface. This fern has been associated with aquatic invertebrates such as freshwater crustaceans (Cladocerans) and often found in Santa María associated with the carnivorous plant *Utricularia* which often captures small crustaceans (F. Ramírez personal observation) (Figure 4E). Less represented plants include the duckweed *Lemna gibba*, the floating primrose-willow (*Ludwigia peploides*) that has yellow flowers, the floating pennywort *Hydrocotyle ranunculoides*, West Indian sponge plant *Limnobium laevigatum*, the swamp smartweed *Polygonum hydropiperoides*

(Figures 4F-G), *Cyperus rufus* (Figure 4H) and the common velvet grass *Holcus lanatus* (Pinilla, 2010).

Santa María del Lago could be considered a biodiversity hotspot within Bogotá. It has been home to 31 bird species (Figures 5A-F) (Molina and Osorio, 1995; Valencia, 1999; Álvarez-León, 2009), but recent estimates report only 11 bird species (Rosselli and Stiles, 2012). Each of these species belongs to a guild as defined by Rosselli and Stiles (2012). This is the case of the duck, *Oxyura jamaicensis*, belongs to the guild of surface swimmers that occasionally dive, and feed mostly on invertebrates and plant material (Rosselli and Stiles, 2012). The zambullidor (diver) *Podilymbus podiceps* belongs to the guild of benthic divers that consume mainly invertebrates (Figure 5A). The cattle egret, *Bubulcus ibis* and striated heron *Butorides striata* and the subspecies endemic to Colombia *Ixobrychus exilis* belong to the guild of wading birds that feed on invertebrates and fish by stalking and striking. The purple gallinule *Porphyrio martinica* and Northern waterthrush *Seiurus noveboracensis* belong to the guild of shoreline walkers that consume invertebrates on mud or bare soil. The endemic species to Colombia the yellow-hooded blackbird *Chrysomus icterocephalus* belongs to the guild of perchers on emergent vegetation that glean for insects or catch them with short flights. Three fish species; the locally known Guapucha (*Grundulus bogotensis*) is an indigenous characin species characterized by conic teeth and no adipose fin (Miles, 1947). It is often found swimming among the floating plants along the water bank where it seeks shelter and feeds on invertebrates. Another fish is El Capitán (*Eremophilus mutisii*) an endemic catfish species to Bogota's Savannah. This species has no ventral fins, feeds on other fish species and grows up to 35 cm (Miles, 1947) making it a delicacy to fishermen but not in Santa María. Another fish is the introduced gold fish (*Carassius auratus*) (Rodríguez-Mahecha *et al.* 2000; Sáenz-Guarín, 2001; Álvarez-León, 2009). The only amphibian, the endemic Andean frog *Hyla labialis* (Álvarez-León, 2009) is less common, and little is known aboutits current status in Santa María. Three mammal species occur in the wetland, among them the introduced guinea pig *Cavia porcellus* (Álvarez-León, 2009). This species is found nibbling at the vegetation but is not easily spotted due to its timid behavior. Most studies regarding the above mentioned animal groups have methodological sampling limitations because they were conducted during only part of the year and sampling techniques were often biased. Therefore, more species are likely to be found. Multiyear research would be advisable especially for most vertebrates with high mobility such as birds.

Phytoplankton recognized as effective nitrogen and phosphorous absorbers, are often used as indicators of pollution in several parts of the world. Within wetlands, the relative abundance of algae can be connected with excess or low nutrient abundance. The main groups of phytoplankton that have been detected in Santa María are represented by diatoms, dinophytes, euglenoids, chlorophytes, and cyanobacteria (Pinilla, 2010). Diatoms (Bacillariophyta) are a distinct group of algae, identifiable under the light microscope by their yellow-brown coloration and by the presence of a thick silica cell wall (Bellinger and Sigee, 2010). Diatoms are often found in the water column or as part of the bottom or on surfaces in Santa Maria. Many of them are used as bioindicators as they are sensitive to pollution. Dinoflagellates (Dinophyta) are mostly biflagellate unicellular algae, although some (e.g. *Stylodinium*) are without flagella and are attached (Bellinger and Sigee, 2010).

Source: Google Earth 2012.

Figure 13. The Juan Amarillo Wetland.

Euglenoid algae (Euglenophyta) are almost entirely unicellular organisms, with a total of 40 genera worldwide (about 900 species) – most of which are freshwater. Cells are typically motile, either via flagella or (in non-flagellate cells) by the ability of the body to change shape (Bellinger and Sigee, 2010). In the freshwater environment, green algae (Chlorophyta) range in size from microscopic unicellular organisms to large macroscopic globular colonies and extensive filamentous growths. This group is commonly found in Santa María near the surface mixed with floating or aquatic vegetation commonly associated with the water fern *Azolla* (F Ramírez personal observation). Blue-green algae (Cyanophyta) occurs widely- throughout freshwater environments ranging in size from unicellular to large colonial mats (Bellinger and Sigee, 2010). These algae are well represented in Santa María by *Anabaena*. Desmids (Desmidaceae) have been found in higher numbers in Santa Maria del Lago compared to other wetland systems in Bogotá (Pinilla 2010). The occurrence of this group of algae is an indicator that the system has fewer minerals, since this group of algae prefers demineralized such as P, Mg, N and K waters (Reynolds, 2006).

Aquatic invertebrates are an important group. According to (Pinilla, 2010), the most abundant aquatic invertebrate group is the dipteran (mosquito larvae) consisting of about 3,000 organisms m^2 during the rainy season and about 2,000 organisms m^2 during the dry season. Other aquatic invertebrates include bivalves (fresh water clams), oligochaetes (worms) and gordioids (Pinilla, 2010). Aquatic invertebrate numbers are low compared to other wetlands. This is possibly a consequence of excess organic matter in the sediment, which increases biological oxygen demand (BOD) and worsens benthic environment conditions for these organisms (Pinilla, 2010). Most of these invertebrates are benthic, and are found borrowing on the bottom mud or near the wetlands substratum.

The question remains open as to the dynamics of other free-swimming invertebrates, such as crustaceans (cladocerans) and many larval stages of beetles and dipterans. To a naturalist, the wetland poses a series of questions about significant relationships among the living taxa. Moreover, aquatic invertebrates are an important element in the food web-dynamics of the wetland. It is known that a number of bird and fish species feed on them. Although food web interactions are important, the ecological balance of this wetland is unknown due to the lack of investigations and governmental intervention. The government has focused mainly on solutions proposed for the local community, but has done meaningless studies at the biological level. There are no studies on the impacts of pedestrians on the local bird populations. Nor are there water quality analyses coupled with assessment of invertebrate-plant interactions or wetland carrying capacity to better understand interactions between humans and the living organisms within the wetland.

Physicochemical Aspects

Information on the physicochemical characteristics of Bogota´s wetlands is scarce, making comparisons with earlier data difficult (Pinilla, 2010). Earlier studies are inconclusive and inconsistent as most of them have only sampled parameters over a limited time range making great variations in physicochemical values difficult to explain (EAAB, 1988; CAR, 1996). To date, the only reliable investigation of physicochemical parameters has been conducted by Pinilla (2010). This investigation provided evidence that the water of Santa María del Lago has few suspended solids and moderate dissolved oxygen concentration of 4.2 mg/L during the rainy season and 5.1 mg/L during the dry season. This moderate oxygen concentration is critical to all life forms as anoxic or oxygen poor conditions provide inadequate conditions for many life forms such as a number of invertebrates and fish. Nitrate and phosphorous levels were low with 0.2 mg/L and 1.6 mg/L respectively. Similarly, nitrites were low 0.002 mg/L and 0.003 mg/L in the rainy and dry seasons, respectively. Low nitrate and nitrite levels limit algal growth and are a good indicator that no sewage or at least very little is being filtered into the wetland. Water temperature during the rainy season was about 18ºC, and 19°C during the dry. Thus, there is very little variation between both seasons' temperatures, but further readings should be recorded on a monthly basis for variation estimation. Total dissolved solids in the rainy season 106 mg/L, compared to the dry 138 mg/L could be considered low when compared to other wetlands that have over 300 mg/L. The amount of solids could influence light penetration, the more suspended solids could block more light and the inverse could be expected with fewer suspended solids. Water transparency in the rainy season measured with a secchi disk was 1.02 m and 1.28 m in the dry. The water transparency is an attribute that is correlated with light penetration that is essential for aquatic plant photosynthesis. During the rainy season, conductivity was 186.9 μS, and conductivity was 211.3 mg/L during the dry season. These are intermediate values when compared to other wetlands of the city. Conductivity measures the concentration of ions present. The more ions could mean more land use by human activities within the water shed, or geological, and biological processes (Roth, 2009). Nutrient status (oligotrophic to eutrophic), pH, water hardness (Ca concentration) and salinity all influence algal communities (Bellinger and Sigee, 2010). Santa María had values during the rainy season of pH 6.8, hardness 100.7 mg/L ($CaCO_3$) and pH 7, hardness 98 mg/L ($CaCO_3$) during the dry season that are known to be beneficial for a number of algal species especially perifitic diatoms. The biological oxygen demand (BOD) was one of the highest among wetlands in the dry season with of 70 mg/L (O_2) compared to only 10 mg/L (O_2) in the dry season. The BOD is a measure of how much oxygen is used-up by microorganisms such as aerobic bacteria in the water. The high levels found in Santa María could indicate a high turnover rate in the decay of plant and animal material.

Environmental Problems

Eutrophication

Eutrophication is the enrichment of water by nutrients, especially nitrogen and phosphorous compounds, causing accelerated algal growth and proliferation of higher forms of plant life to produce an undesirable disturbance to the balance of organisms and the quality of the water supporting them (Dokulil and Teubner, 2011). Santa María del Lago wetland was evaluated according to two indices, the index of biological communities (BI) and limnological conditions index: (LICOI) applied and proposed by Pinilla (2010).

According to these indexes, Santa Maria has an intermediate state of deterioration based on limnological conditions (Pinilla, 2010). Visit to the wetland indicate that eutrophication might be causing water conditions to produce noxious odors in portions of the wetland (F. Ramírez personal observation).

Remediation Efforts

The local community and Bogota's city office (former DAMA) have combined efforts to re-forest the wetland with trees (Moreno *et al.*, 2005). Bogota's city office also began ecological restoration actions which have been aimed to improve environmental quality. The local community together with other institutions built a sewage removal connector system. This system collects the incoming sewage water and diverts it into the city's sewer system (Moreno *et al.*, 2005).

Laws, Action Plans and Conservation

In 2003, the Bogota's office wrote the Action Plan. Recently, a law passed (Resolución 7773 of 2010) that adjusted and approved the environmental action plan of Santa María de Lago. The local community has played an important role in conservation. This wetland welcomes hundreds of school and university student sannually who are involved in wetland conservation education.

Cordoba Wetland

General Information

The Cordoba wetland, located in the northwestern section of Bogotá, has experienced land fragmentation over the past. This phenomenon has been caused by highway construction, namely the Suba highway, Cordoba Avenue 116thand 127thstreets.

These roads cross or, in particular cases, interrupt the wetland at points. Suba highway transverses the wetland and has caused the wetland to be divided into two main portions (Figure 6). Cordoba wetland has been divided into three portions, one above Suba highway which has not been fenced, the fenced portion below Suba highway and an area near 127th street that has almost disappeared due to urbanization. Fencing has been controversial due to lack of entrance control and patrol.

The local community's security complaints impeded fencing on the portion above Suba highway. Their argument was that fencing without any entrance or without security would leave visitors to the wetland vulnerable to thieves. On the other hand, Bogotá's Water Company argument for fencing was security and to prevent vagrant homesteading on the property. There is concern because fencing offers no control to the garbage disposal, another environmental issue within the wetland.

The Cordoba wetland covers 14.73 Ha. (Rosselli and Stiles, 2012) much of its original land has been used for housing, commercial development and even club settlements (Club Choquenzá and Club de los Lagartos) (Figure 6). Cordoba is immersed within neighborhood blocks of Niza-Córdoba, Niza VIII and Lagos de Córdoba. Since the 1960s, the Cordoba wetland was used for agricultural purposes for production of cattle, corn and potatoes (Palacio and Van der Hammen, 2007).

Today the wetland has no agricultural-related activities but faces serious environmental issues such as sewage inputs, solid residue disposal, vagrancy, eutrophication and biodiversity loss. Although the wetland's perimeter has been fenced with management and biological signs having been posted (Figures 7A and B), the wetland faces serious problems by entrance of a series of channels that input sewage-rich waters and from the Juan Amarillo River (Figures 8A-D).

These characterize Cordoba as a type A hydric flow system, where the original inflow and outflow channels remain intact with a high water flow volume (Van der Hammen *et al.*, 2008). These outflow channels connect with Bogotá's River basin. People often visit the wetland. Within the area below Suba Highway, some unpaved trails are used by bikers, runners and joggers. These are common visitors especially on weekends. Other, less concerned visitors dine near the water´s edge as evidenced by the great amount of plastic cups, plates and cutlery left behind as a reminder of human intervention. There are even individuals that use the wetland as a place to use narcotics and illicit substances.

Source: Google Earth 2012.

Figure 14. La Vaca Wetland.

Not all aspects are negative. There are naturalists that are attracted by the bird population, plant diversity and general wetland environment. More education is clearly required in order to consolidate a real wetland appreciation by the local city resident. Perhaps the international community might be more interested and could view this wetland as one of the last of these unique places.

Biodiversity

The perimeter of the upper portion above Suba highway has a well-established tree community with many native and introduced species. Introduced species include *Eucalyptus*, *Acacia*, pines, and urapan (*Fraxinus chinensis*) a six to seven meter high tree (EAAB and CIC, 2000). These trees have been planted unevenly on the outer margin of the wetland. Native species above and below Suba highway include the alder (*Alnus acuminata*), Caucho sabanero (*Ficus soatensis*), willow *Salix humboldtiana,* etc (Moreno *et al*., 2005). Tree species such as the willow (*Salix humboldtiana*) grow near the water's margin or within the water. Most other tree species are found on solid ground surrounding the wetland's perimeter. It should be noted that a consistent study about indigenous and introduced tree species is required to understand the taxonomic status of Cordoba.

Birds are represented by 18 species (Rosselli and Stiles, 2012). These belong to different guilds according to Rosselli and Stiles (2012). In the case of ducks, *Dendrocygna autumnalis*, *Oxyura jamaicensis* and the migratory species *Anas discors*, belong to the guild of surface swimmers that feed on invertebrates. *Podilymbus podiceps* (see gild in the Santa María section). The little blue heron *Egretta caerulea*, cattle egret *Bubulcus ibis*, striated heron *Butorides striata*, the great egret *Ardea alba* belong to the guild of waders-waiters. The American purple gallinule *Porphyrio martinica* and the common moorhen *Gallinula chloropus* belong to the guild of walkers on aquatic vegetation. The spot-flanked gallinule *Gallinula melanops*, the American coot *Fulica Americana*, belong to the guild of surface swimmers. The lesser yellowlegs *Tringa flavipes*, greater yellowlegs *Tringa melanoleuca*, solitary sandpiper *Tringa solitaria*, spotted sandpiper *Actitis macularius* and the northern waterthrush, *Seiurus noveboracensis*, are migratory except for the southern lapwing *Vanellus chilensis* and altogether belong to the guild of shoreline walkers that consume invertebrates in mud or bare soil. The torrent tyrannulet *Serpophaga cinerea*, the endemic species to Colombia Apolinar' s wren *Cistothorus apolinari* and the yellow-hooded blackbird *Chrysomus icterocephalus* belong to the guild of perchers on emergent vegetation that search for insects or catch them with short flights (Rosselli and Stiles, 2012). Aquatic plants form aggregations of various species along the wetlands banks on the portion above Suba highway (Figures 9A-F). This portion has a remnant-flow channel that has been filled with garbage. On the portion below Suba highway, aquatic plant communities have spread over the water surface and cover most of the wetlands water-habitat. This portion of the wetland has experienced lower water inputs that negatively impact aquatic plant communities because of exposure to dry conditions (F. Ramírez personal observation) (Figures 10A-D). Aquatic plants are represented by Duckweed *Spirodela intermedia* a floating aquatic plant that covers most areas near the water´s edge and grows carpet-like on the water´s surface, the rush *Juncus* sp., that has long stems,the smooth beggarticks *Bidens laevis* (Figures 9F and 10D), floating marsh pennywort *Hydrocotile ranunculoides, Ludwigia peploides* with yellow flowers and *Thypha latifolia* with long leaves visible from the distance (Figure 9A) (EAAB and CIC, 2000; Moreno *et al*., 2005). Again, there are no studies that record in detail the plants found within Cordoba and many of the cited texts have misidentifications.

Physicochemical Aspects

Few physicochemical studies have been conducted in the Cordoba wetland. Of these, most have been carried out by the government but have methodological and analytical inconsistencies.

Environmental Problems

Sewage

Since the 90s, the Cordoba wetland has been impacted by sewage discharge (Palacio and Van Der Hammen, 2007). It receives the discharge via three different channels, namely Cordoba, del Norte and Los Molinos (Figures 8 A-D) (Moreno *et al*, 2005). The del Norte channel receives runoff waters, other inflow channel waters such as Contador and Callejas, also inflow from several streams (Contador, Bosque de Medina, Gimnasio Femenino, Trujillo and Delicias del Carmen). Los Molinos Channel receives polluted waters from the nearby neighborhoods and from five streams (Sagrado Corazón, Gimnasio

Los Cerros, Santa Ana, La Chorrera de Molinos y el Pedregal). Sewage produced by the nearby neighborhood enters the wetland (Moreno *et al*, 2005). The water outflow from this wetland is connected to the main Bogotá River, which has a history of severe pollution over decades. The wetland lacks a hydraulic connectivity system with the city's sewer system to divert polluted waters, and at its highest point there is little water inflow (Moreno *et al*, 2005). Pollution is a critical issue and is seriously threatening this wetland. Runoff water also drains into the channels and wetlands and is considered a pollution source for Cordoba. Bogotá's Water Company has taken no measures to improve the environmental conditions of the wetland, and there is little commitment by other governmental organizations such as the Bogotá's city office.

Solid Waste

People dispose their garbage around the wetland. It has also been observed that the area above Suba highway, just over the water's edge has been turned into a solid waste dump where large quantities of garbage and debris are disposed illegally (Figure 11). In the area below Suba highway, solid dumping is less severe and the main issue is the garbage left by vagrants and often visitors that have little environmental commitment.

There is no control over people who dump in the wetland making solid disposal control difficult although the government has proposed several action plans.

Eutriophication

Cordoba faces a great eutrophication problem. Reduced freshwater entry into the wetland from the near-by channels and the dominance of sediment-rich waters has increased the problem.

Once the wetland is severely impacted by eutrophication the area losses its hydric balance and becomes clogged. After this process occurs, the land, nearly dried becomes available for illegal landfillers to justify urbanization or commercial projects. These illegal landfillers act on behalf of corrupt politicians that only want to benefit from a share a share of the construction projects. Few investigations have focused on these issues that demand action from non-corrupt policies and well-structured environmental knowledge. Perhaps future generations might start a change in this scenario.

Restoration Efforts

In the 1990s the local community developed restoration and conservation efforts (Palacio and Van Der Hammen, 2007): however, most of this information is not available to the general public and few actions have been initiated. Today, Bogota's Water Company and the government have abandoned the wetland. This is clearly observable by the way sewage, garbage, security and water flow/outflow is managed. The wetland has clear potential opportunities to enhance its environmental condition, such as construction of upstream sewage treatment plants. Although these measures are urgently needed, the government seems to conduct its actions at its own pace.

Source: Google Earth 2012.

Figure 15. El Burro Wetland.

Source: Google Earth 2012.

Figure 16. The Capellanía Wetland.

Source: Google Earth 2012.

Figure 17. The Jaboque Wetland.

Laws, Action Plans and Conservation

Action plans for the Cordoba wetland have been controversial, many of which have not been approved. There was a management environmental plan proposed in 1996 by Ecology and Environment Inc and Hidromecanicas L.T.D.A. in association with Bogotá's Water Company. This proposal was not approved due to inconsistencies. Some time passed until Bogota's office proposed the management plan for Cordoba in 2008 (SDA, 2008). The plan provides a diagnosis of physical, ecological, social and cultural wetland associated aspects to solve environmental problems that include sewage input, a deficient hydraulic system, ecological isolation, wetland fragmentation, reduced water flow, solid waste and debris disposal, invasive species, and lack of security (SDA, 2008). The plan also proposes management zones. These zones include preservation, conservation and control. The plan incorporates five programs 1) applied and basic research, 2) education and communication with the local community, 3) remediation, 4) management and sustainable use and 5) governmental management (SDA, 2008). The plan has an important social component and seeks to work with the local community. The management plan made agreements with the environmental studies institute (IDEA) ascribed to Universidad Nacional de Colombia and Bogotá's Water Company. The outcome of this action plan has been negligible, most of the plan remains on paper and actions are awaiting. One of the current problems is the bureaucracy level (huge number of papers and permits to conduct actions, and long periods of time) and corruption. Just by visiting the wetland the local resident gets a grip on the poor environmental conditions within Cordoba.

Future actions would need to be enforced by the local community. It is hoped that the wetland is not completely forgotten and left in corrupted hands.

La Conejera Wetland

General Information

The Conejera wetland dates back to indigenous times when the Muiscas were among the first settlers to cultivate the land with corn, potatoes, endemic tuber species (cubios) and yucca (DAMA, 2000). After this indigenous period and during the Spaniard conquest, the wetland was part of a ranch (locally known as Hacienda La Conejera) owned by the Jesuits during the17th century. This ranch was later sold in 1775. Later on the 20th century, the wetland started as a series of interventions, such as landfilling that caused a substantial reduction in size, tree logging, and cattle paddocks that eventually turned into urbanized areas (DAMA, 2000). Although there are accounts of reduction from the original size, there is no evidence documenting the actual number of hectares lost. In 1971 the wetland was divided into cattle ranches, but by the late 80s the first urbanization began being constructed robbing the wetlands of its original lands. Today, the remaining wetland is part of Suba, a locality toward the north western area of Bogota's savannah. This wetland is limited by a road that connects the two villages, Suba and Cota, to the north. It borders Suba´s hills to the east, and merges into the Juan Amarillo River to the south. The wetland's outer boundaries are urbanized by Compartir with city-blocks zoned mainly for housing within the Suba locality (Figure 12).

The wetland is not an isolated system and thus, receives input waters from its main supplier, Salitrosa stream (also known as Conejera stream), and underwater springs. Therefore, La Conejera wetland is characterized as Type A hydric flow system, wherein the original inflow and outflow channels remain intact with a high water flow volume (Van der Hammen *et al.*, 2008). Much of the outflow is connected to the Juan Amarillo River that then connects to the Bogotá River basin.

La Conejera wetland covers 21.85 Ha. (Rosselli and Stiles, 2012). The wetland has been fenced in an effort to limit possible landfilling operations and to protect the wetland's biodiversity. The local community and the non-governmental organization (NGO), Fundación Humedal la Conejera, has developed the main strategy for conservation and environmental protection of the wetland. Currently this (NGO) has provided a number of developmental actions on topics such as environmental education, restoration and policy. It has provided a series of plans and complementary actions to improve the current environmental issues within the wetland. Although there are several positive points, the wetlands are viewed by police as a ' dangerous place where visitors wander at their own risk' .

Biodiversity

La Conejera wetland has several times been reforested with native species. The land near the wetland is currently planted with diverse tree species for instance, the endemic *Polymnia pyramidalis* that can reach up to 10 m and is considered a fast growing species ideal for restoration, *Alnus acuminata* which grows to 30 m in height and willow *Salix humboldtiana*, from which uric acid is extracted from cork for medicinal applications (EAAB and CIC, 2000). Trees representing different native genera, such as *Alnus, Sambucus*, *Abutilon*, *Salix*, and introduced such as *Senna Prunus*, *Bocconia*, *Eucalyptus*, *Acacia*, *Solanum* and *Pittosporum* have been planted around the wetlands' drier areas (Moreno *et al.*, 2005). Although the NGO Fundación La Conejera has been generating environmental solutions there are no published documents available to the public for study of the trees from this wetland.

Birds reported from La Conejera wetland area include 24 species (Rosselli and Stiles, 2012). The bird species and guilds found in La Conejera are generally the same as those discussed in the previous sections of this chapter. The ducks, *Anas discors*, *Nomonyx dominicus*, and *Oxyura jamaicensis*, the diver *Podilymbus podiceps*, the egrets *Egretta caerulea*, *Bubulcus ibis*, *Butorides striata*, *Ardea alba*, *Nycticorax nycticorax*, the bare-faced ibis *Phimosus infuscatus* belong to the guild of waders in search for invertebrates. The gallinule *Rallus semiplumbeus*, sora rail *Porzana Carolina* is from the guild of walkers on aquatic vegetation that consume invertebrates and plants, the purple gallinule *Porphyrio martinica* that is often observed on floating vegetation and rarely swimming (Asociación Bogotana de Ornitología, 2000), the gallinules *Gallinula chloropus*, *Gallinula melanops*, the coot *Fulica americana,* which are frequent among the wetlands bank among the aquatic vegetation, *Vanellus chilensis*, the noble snipe *Gallinago nobilis* frequents the banks in search of food in the mud with its long beck. The lesser yellowlegs *Tringa flavipes* and the solitary sandpiper *Tringa solitaria* are migrants that nest in North America (Asociación Bogotana de Ornitología, 2000).

The torrent thyrannulet *Serpophaga cinerea* which is found near the water ready to capture insects that flay among the vegetation, Apollinar' s wren *Cistothorus apolinari* lives within *Juncus* sp, *Typha* and among vegetation, the northern waterthrush *Seiurus noveboracensis*a migrant from North America and the yellow-hooded blackbird *Chrysomus icterocephalus*. Dogs have caused serious attrition of the wetland birds, such as *I. exilis,* which has been extirpated from this wetland (Rosselli and Stiles, 2012).

Mammals in this wetland include the gunea pig *Cavia porcelllus*, which can be easily be spotted among the vegetation where it seeks cover, the long-tailed weasel *Mustela frenata* which is found near the water known to tolerate human presence, the red-tailed squirrel *Sciurus granatensis*, often actively searching for nuts and other fruits, and Thomas' small-eared shrew *Cryototis thomasi*, which is found in Bogotá at 2600 m to 3500m. Amphibians include the indigenous frog *Hyla labialis* that shows a great color variation, ranging in size from four to seven cm in length. Fish include the introduced common carp

(*Cyprinus carpio*) which is an omnivore. Thirty six species of invertebrates have been found, but this information is not available to the public (EAAB and CIC, 2000; Asociación Bogotana de Ornitología, 2000).

The aquatic plant community is dominated by totora grass, *Scirpus californicus*, that forms large patches near the water's edge and on isolated points within the wetland.

Other plants include the wide spread yellow-flowered *Bidens laevis*, floating primrose-willow, *Ludwigia peploides*, and the endemic species, *Senecio carbonelli*, (EAAB and CIC, 2000). This species was considered extinct (Institutio Alexander Von Humboldt, 1997), but in 1998 it was rediscovered in the wetland by Fundacion Humedal la Conejera. Since then, neither the status nor biology have been studied or evaluated. The available aquatic plant information should be handled with caution because the cited reports are relatively old, incomplete, and recent updates are required. Moreover, there are no accounts of systematic or consistent studies concerning aquatic plants at La Conejera.

Environmental Problems

Sewage

Near the perimeter of La Conejera wetland, Compartir urbanization, a housing complex, was the main sewage input source into the wetland in the early 90s. Likewise, nearly 15 neighborhoods poured over five billion cubic meters of sewage into the wetland in the 90s. This problem was solved by Bogotá's Water Company by redirecting it into Bogotá's sewage system. In 2005, raw sewage entered also as a result of discharge by a Hospital (Clinica Corpas) and from the nearby neighborhoods, Londres and Las Acacias (Moreno *et al.*, 2005). Although there have been some efforts to solve these sewage issues, La Conejera still receives sewage inputs from neighborhoods and illegal urbanizations.

Landfilling

Landfilling has been one of the major problems the wetland has faced over the past mainly over the late 80s and 90s (DAMA, 2000). Landfilling began in 1993 with the Compartir urbanization. Landfills were made mainly in part of the eastern portion of the wetland (DAMA, 2000; Moreno *et al.*, 2005). This caused Salitrosa stream to change its course. Today due to the management and fencing of La Conejera, landfilling has stopped and seems a past issue.

Restoration Efforts

Community involvement in restoration and environmental protection is key for wetland conservation. Since 1993, Fundación Humedal La Conejera, has taken actions to defend, conserve and restore La Conejera Wetland. Some of the main achievements of this NGO have been to: stop land-filling, fence the wetland's perimeter, plant trees near the wetland, improve the sewage system, create ecological connectivity, rediscover the endemic plant species, *Senecio carbonelli*, and restore the inflowing stream (Salitrosa stream). This NGO has worked together with the local community and Bogota's Water Company to restore the wetland's ecological functions.

Community involvement includes education and activities involving boys and girls and teenagers, as well as school teachers. For more information see http://www.fhlc.co/

Laws, Action Plans and Conservation

Fundación Humedal la Conejera has been the main entity implementing legal actions in the wetland. In association with Cundinamarca' s state governmental corporation (CAR), it proposed a management plan in 1998 (FHLC, 1998).

The accomplishments of this plan had little impact and had major inconsistencies and thus, due to these issues it will not be discussed herein. Another management plan was proposed in 2006 by Fundación La Conejera and Bogota's Water Company.

This plan was structured to define ecological zones within the wetland. These include conservation, restoration, environmental education and entertainment zones. The plan conveys a similar structure to that of the Cordoba wetland (2008) (See: laws, action plans and conservation under Cordoba wetland).

The plan has been successful according to Fundación Humedal de Cordoba, but there remains more to be done. The wetland still faces sewage issues and there is little-in-depth-information to the public about reports on restoration, and general management aspects which make it difficult to assess actions. This NGO also proposed an ecotouristic plan for la Conejera, but it is not available to the general public.

Source: Google Earth 2012.

Figure 18. The Techo Wetland.

JUAN AMARILLO WETLAND

General Information

Juan Amarillo wetland is the final remnant of a great lake that existed over Bogotá. The wetland is also named Tibabuyes meaning ' land of farmers' in the original Chibcha language (DAMA, 2000). The Chibchas were the indigenous culture established on the area before the Spaniard conquest. The Juan Amarillo River wetland receives its name from its water connections to the Juan Amarillo River (Figure 13). Prior to 1956, the wetland had little human intervention. Juan Amarillo River was diverted from its original course in 1969 by the construction of a levee (Jarillón), which caused water loss and drier conditions because water could no longer flow into the wetland. This caused a major water loss to about 65% of the original volume and had negative effects on the aquatic vegetation, which was reduced to over 30% of its original size (DAMA, 2000). It should be noted that no other environmental effects were measured and this information provides only general outcomes. The levee caused large areas within the wetland to dry-up becoming pasturelands in 1977. This became an opportunity for landfills to occur and aggravate the initial problem even more. Illegal landholders, began building small houses, and in the 80s the land was sold to larger landholders that used the terrain to build large urban communities, such as Colsubsidio (DAMA, 2000). Urbanization continued, and large neighborhoods were built on the nearby wetland zones, these comprise Bachue, Carmen and Tibabuyes as well as roads.

A water treatment plant was built in 2000 by the Suisse Company Suez Lyonnaise de Eaux Degrémont. Later in 2004 the plant's concession was passed to Bogotá's Water Company, which currently operates it. Currently, the wetland has been divided into three sectors- upper, mid and lower. The upper section is characterized by a concrete bank built by Bogotá's Water Company. The structure stands like a wall on the water´s edge. Plans were not open to public input prior to construction, and it has received much criticism because it did not take into account environmental parameters, such as gradual slopes that would enable plants and aquatic organisms to colonize the water's edges. The upper section also has a sidewalk for pedestrians along the water bank. Water from the Salitre River flow into the wetland within the mid-section. This section has been less intervened than the upper section, and thus, has no concrete bank. The lower section displays a well-established plant community and less human intervention.

The Juan Amarillo wetland area covers a total of 120.66 Ha. (Rosselli and Stiles, 2012). It is characterized as a Type A hydric flow system, where the original inflow and outflow channels remain, and there is a high water flow volume (Van der Hammen *et al.*, 2008). This wetland receives inflow waters from the Negro and Salitre rivers and drains into the Bogotá river basin. It is found in the northwest section of Bogotá. Its limits to the west is the Bogota River, to the east is the Santa Cecilia community and Lisboa community and to the north and south with other neighborhoods.

Biodiversity

Landscape configuration of the Juan Amarillo wetland has changed dramatically over the past 55 years. Man-made structures such as the concrete bank discussed above, land-filling operations, sewage and solid disposal problems have affected the wetlands biodiversity. Invasive species have also been a major problem, this is the case of the invasive grass species, *Pennisetum clandestinum*, which has increased in abundance since the 80s and is now threatening the survival of native species (Bejarano and Bonilla, 2009).

The northern and southern parts of the wetland are home to 37 tree and 17 herbaceous plant species (Guzman and Flórez, 2005). However, this information remains to be confirmed. Trees found along the wetland's perimeter in the dryer areas comprise introduced species, such as *Acacia* sp, *Eucalyptus* sp, *Pinus* sp, *Prunus* sp. and *Fraxinus chinenesis*. *Eucalyptus* and *Pinus* have adverse effects on soil chemical and biological properties; because these species often add chemical substances that kill other plant species and also use large amounts of water. Native tree species include *Alnus acuminata, Salix humboldtiana,* the yellow flowered *Abatia parviflora* which grows up to a height of 20 m, *Escallonia* sp, a wind resistant shrub, *Adipera tomentosa, Ficus* sp, etc.

The Juan Amarillo wetland is home to 21 bird species (Rosselli and Stiles, 2012). Many of these are also found in the other wetlands of the city. It remains to be determined if local and endemic birds fly among wetlands and how these interactions have been impacted by the degrading state of most of the areas.

Mammals include the guinea pig, *Cavia porcellus* that live in groups up to 10 individuals. Reptiles include two snakes, the thickhead ground snake, *Atractus crassicaudatus* that can reach up to 40 cm and is often active during the day. It can be distinguished by its black-colored dorsal region and yellow-colored ventral region. This snake is reportedly common in areas around Bogotá. The other snake is the night ground snake *Liophis epinephelus bimaculatus* which reaches over 47 cm and is often green-colored. Amphibians include the frog, *Hyla labialis*. Although, there is a considerable water pollution, the fish commonly known as guapucha, *Grundulus bogotesis* was observed in the upper section of the Juan Amarillo wetland in 2004 (Ramírez, F personal observation). Evaluation of population dynamics and the current status of these vertebrates within Juan Amarillo are unknown.

Invertebrates, such as planarians are an abundant group with about 18,000 organisms m^{-2} found during the rainy season and 19,000 organisms m^{-2} counted during the dry season (Pinilla, 2010). Some invertebrate groups found in Juan Amarillo are indicators of contamination by organic matter, for example worms (Oligochaeta) and mosquito larvae (dipterans). In contrast, the wetland lacks organisms typically found only in clean waters for example stoneflies (plecoptera), aquatic mites, hydras, and bivalves (Bivalvia) (Pinilla, 2010).

Source: Google Earth 2012.

Figure 19. The Tibanica Wetland.

There are 17 species of aquatic plants found in the Juan Amarillo wetland (Pinilla, 2010). The relative abundance of these plants was estimated by frequency of occurrence (FO see Santa María section above). The most abundant aquatic plant was *Hydrocotyle ranunculoides,* which seems to have spread throughout the wetland's water banks. Another greatly abundant plant is duckweed *Spirodela intermedia* (FO=89) a floating plant that benefits from the high nitrogen levels found within the wetland and that will be discussed on the next section. Other floating plants seem less adapted to this highly polluted environment, for example the least duckweed, *Lemna minuta* (FO= 5.3), *Lemna minor* (FO= 10.12), and *Lemna gibba* (FO= 2.5). The smooth beggarticks *Bidens laevis,* (FO= 67), is an abundant rooted species found along the banks. Other mud-rooted plants, such as *Solanum americanum* (FO= 6.62), the swamp smartweed, *Polygonum hydropiperoides, Rumex conglomeratus* (FO= 0.62), a terrestrial plant that invades areas near the water, and *Cyperus rufus* (FO= 1.25) are common. The introduced kikuyu grass,

Pennisetum clandestinum (FO= 34.5), a terrestrial grass species and the common velvet grass *Holcus lanatus* (FO= 10.75) appear to thrive in the water of this wetland, but *Schoenoplectus californicus* (FO= 3.38) seems to be negatively influenced by the high pollutant content. The introduced water hyacinth, *Eichornia crassipes* (FO= 68.37) is a floating plant that seems to take advantage of the altered conditions to spread over the water surface. The one-row yellowcress, *Nasturtium officinale* (FO= 1.25), and *Rorippa pinnata* (FO= 5.87) have low abundances (Pinilla, 2010). Most of these plants are common throughout Juan Amarillo and other wetlands within the city. None of them have a category of threatened or endangered within the Juan Amarillo wetland but much more in depth research is required to determine to what extent this holds true.

Phytoplankton is represented by diatoms, dinophytes, euglenoids, chlorophytes, and cyanobacteria (Pinilla, 2010). Dominant groups included euglenoids and chlorophytes (Pinilla, 2010). *Cyclotella* is the most abundant genus in Juan Amarillo during the rainy season (Pinilla, 2010).

Physicochemycal Aspects

Pinilla (2010) measured a number of physicochemical variables in the waters of the Juan Amarillo wetland during the wet and dry seasons. Although this wetland was qualified as having ' regular limnological conditions' , it was the richest in nutrients (ammonium, 10 mg/L during the rainy season and 14 mg/L during the dry season season and orthophosphate levels were 1.1 mg/mL during rainy and 9.9 mg/mL during dry seasons), indicating a greater degree of eutrophication when compared to wetlands such as Santa María del Lago, Jaboque, Tibanica Guaimaral and Córdoba (Pinilla, 2010). During the rainy season, measured variables included water temperature, 20.3°C, which is higher than in other wetlands except for Tibanica; total dissolved solids, 127 mg/L, which was lower than in other wetlands and could indicate a high inflow rate and conductivity, 214 μS, a similar value when compared to other wetlands within the city. Dissolved oxygen was 0.15 mg/L, an extremely low value, which is already detrimental to aquatic life and symptomatic of the high level of eutrophication, and pH was 7.7, which means that pollutants such as detergents could be increasing the pH toward more alkaline levels. Chloride concentration was 20 mg/L a low value found among wetlands. Water hardnessat 69 mg/L ($CaCO_3$) was a relatively moderate level, and secchi disk transparency was 0.63 m which is among the lowest values for the examined city' s wetlands. The levels of suspended solids was 20.66 mg/L, a low value, whereas nitrate concentration was among the highest values found among wetlands at 0.03 mg/L. Nitrite levels were 0.024 mg/L, an intermediate to high value, and BOD was 84 mg/L O_2, a high value meaning increased bacterial activity.

Variables measured during the dry season included water temperature at 16.5°C which was lower than in the rainy season, total dissolved solids at 408 mg/L, a high value possibly as a result of less water entering into the wetland during the dry period, and conductivity at 585 µS, an intermediate level compared to other wetlands. Dissolved oxygen was 1.3 mg/L, which was low and often below that required by fish, and pH was neutral at 7.06. The chloride concentration was 50 mg/L, a moderate level compared to other wetlands. Water hardness was 136 mg/L ($CaCO_3$), a relatively moderate to high level, and secchi disk transparency was 0.29 m, which was lower than during the rainy season resulting in greater inhibition of light penetration to lower water strata to hamper photosynthesis. Suspended solids were high at 52.38 mg/L contributing to the decreased water transparency indicated by the secchi disk reading. Nitrate levels were high at 0.3 mg/L. Nitrites were an extremely high level at 7 mg/L indicative of major pollutant input, and BOD was 25 mg/L O_2, one of the highest values for the wetlands studied.

Environmental Problems

Sewage

The upper part of the Juan Amarillo wetland was modified to construct a water treatment plant utilizing sediment-pools to clarify water flowing from the Juan Amarillo River in 1969 (Bejarano and Bonilla, 2009). These pools allow the water to decant or settle as part of water treatment to remove dissolved solids.

The water plant and the water-treatment pools were abandoned due to corruption and monetary issues from 1977 to 2000, when Bogota's Water Company, began a wetland recuperation program. The program mainly focused on re-opening the water treatment plant and particularly the water-sedimentation pools for dissolved solids removal. The upper portion of the wetland also receives underground sewage drainage causing major environmental issues (Shmidth-Mumm, 1998). Dredging the bottom of the upper portion of Juan Amarillo has been carried-out on several occasions (Herrera *et al.*, 2004). Dredging is done as a means to remove the high amount of solids that pile-op on the wetlands bottom as a result of the high amount of sewage and residues from construction that clog water flow. The current state of this wetland is bad.

Source: Google Earth 2012.

Figure 20. Part of the Guaymaral-Torca Wetland near the North Highway.

Although a water plant is currently operating, the negative effects of high nitrogen substances (ammonium, nitrate and nitrites as well as phosphates) have impacts on aquatic life, making it difficult for vertebrate species and oxygen dependent invertebrate species to thrive within the polluted waters. More actions aimed at improving the aquatic environment are clearly required.

Landfilling

During the 90s landfilling was the main mechanism for illegal urbanization establishment (Bejarano and Bonilla, 2009).

Figure 21. Aquatic plants within the Guaymaral-Torca Wetland. A) Plant community mainly *Cyperus*. B) *Polygonum hydropiperoides* among grass. C) invasive grass. D) *Hydrocotyle ranunculoides.*

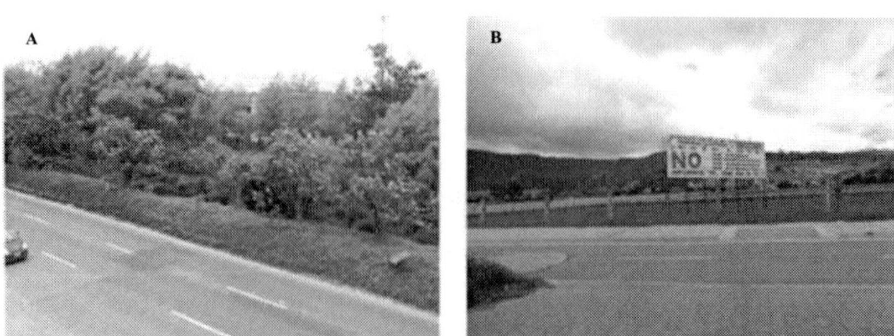

Photos by Fernando Ramírez.

Figure 22. North Highway (Autpista Norte) has severely impacted this wetland. A) View from bridge. B) Nearby areas to the highway have been lost by land filling.

Total area reduction occurred as a consequence of construction of local communities, such as Santa Cecilia, Luis Carlos Galán and El Rincón (Herrera *et al.*, 2004).

Eutrophication

High amounts of dissolved salts and the already mentioned high level of nitrogen compounds as a result of the high raw sewage discharge and inflowing waters that contain sand residues have accounted for an increased amount of eutrophication. For example, neighborhoods such as Lisboa, often pump raw sewage water into the wetland. According to the biological index (%BIs) and the index of limnological conditions (LICOI) applied by Pinilla (2010), the Juan Amarillo wetland is in an intermediate state of deterioration. Moreover, high sediment inflow along with a reduction in the amount of water flowing into the wetland as a result of the levee built to modify the normal course of the Juan Amarillo River, has been one of the major contributors to the demise of the Juan Amarillo wetland (Herrera *et al.*, 2004). About 3,600 mg/L of dissolved solids have been detected at the wetland´s entrance and about 2,170 mg/L remain inside the wetland (Herrera *et al.*, 2004).

Restoration Efforts

Restoration efforts by Bogota's Water Company and the Bogotá's city office are being planned to re-establish wetland boundaries by regaining space lost by illegal housing (Herrera *et al.*, 2004). Restoration efforts started from 1989 to 1995. Restoration actions have been controversial because they have modified habitats within the upper portion of the wetland (see above section about the levee) and the water treatment plant is currently insufficient, and raw sewage continues to inflow into the wetland. The NGO, Fundación Alma Tibabuyes has been involved in the environmental management of Juan Amarillo. This NGO has produced educational material in the form of leaflets that inform the general public about the importance of the wetland, pointing out its three zones upper, mid and lower, general bird and plant populations, biological information and appropriate ecologically-friendly attitudes that visitors should have when visiting the wetland. It encourages wetland visitors to use appropriate garbage disposal, not to feed aquatic birds, not to run or use bicycles or motorcycles within the wetland, prohibits hunting and fishing, prohibits alcohol and drugs usage and denounces sewage inputs. Although this wetland is open to the general public, Juan Amarillo is immersed within a problematic locality because nearby neighborhoods have very low incomes and people are often forced to become thieves. Police recommend that is best not to venture into Juan Amarillo wetland due to possible life threatening situations.

Bogota's Water Company attempted to restore the Juan Amarillo wetland in 2002 by creating a shallow (2–4 m), 22-ha lake free of major aquatic plants (Zapata-Anzola *et al.*, 2009). During the basin reconstruction, nutrient-rich sediments provoked a high trophic status that caused appearance of major algal blooms (Zapata-Anzola *et al.*, 2009). Thes ealgal blooms, mainly composed of the blue-green *Woronichinia* cf. *fusca* (Cyanophyceae) have been controlled by potassium chloride (Rivera-Rondon *et al.*, 2008). Rice shoots have also been used to control blue-green algal blooms; however, this causes oxygen depletion due to bacteria that grow along the rice stems (Rivera-Rondon *et al.*, 2008). These methods only partially solve the problem. The best way will be to control the amount of nitrogen entering the systemby controlling pollution at its source.

Laws, Action Plans and Conservation

Bogotá's city office and Bogota's Water Company proposed a management plan for the Juan Amarillo wetland in 2012. The plan provides a general diagnosis of the wetlands environmental problems such as hydric alterations, sewage contamination of the wetland, poor sewage conductivity, garbage disposal, debris waste, reduced wildlife habitats, aquatic biodiversity loss, invasive species, grazing mammals such as cows and horses. Other issues include noise pollution, few environmental protection plans and actions, low social income, un-planed recycling near the wetland, soil-use conflicts, new housing projects and illegal land conflicts (SDA, 2010a).

The plan establishes different zones such as ecological restoration and recuperation zones, a management zone, and a terrestrial zone. The plan has five strategies common among all wetlands (See Laws, Action plans and Conservation on the other above sections of this chapter). However, the plan does not establish further actions that would lead to environmental protection and law enforcement (SDA, 2010a). The area currently holds the status of ecological park (Herrera *et al*, 2004).

Source: Google Earth 2012.

Figure 23. The Meandro del Say Wetland.

La Vaca Wetland

General Information

Immersed within the Kennedy locality, on the southeastern side of Bogotá, la Vaca wetland covers 7.9 (DAMA, 2006) to 40 Ha (Pinilla, 2010) (Figure 14). This wetland is immersed in one of the most populated areas within the city. La Vaca is also known by the name Techovita after a native indian chief that lived in the area. This wetland originated when the large Tintal Lake was divided by the construction of Techo' s airport and Las Américas highway in 1930 into five portions, and one of them became La Vaca. Agriculture was the main economic activity during the 1950s. The 70s marked the beginning of illegal landfills caused by illegal immigrants that begin to homestead the land for their own benefit. On the 80s, migrant people from the countryside established themselves around and within the wetland mainly by landfilling operations (Moreno *et al.*, 2005). Furthermore, in the 90s the landfilling tendency continued. Currently, la Vaca is divided into two sections (A and B) that are split by a neighborhood portion. La Vaca is not fed by external streams or rivers; it is a system with limited water flow mainly as a result of rainwater runoff (Van der Hammen *et al.*, 2008). By the year 2000, this wetland received raw sewage from man-made channels (DAMA, 2000). Sewage is still flowing into the wetland making life conditions difficult.

Biodiversity

The wetland has a long history of deterioration. As occurs with other wetlands, La Vaca wetland has been poorly monitored and biodiversity has diminished to very low levels, to the point that only a few original aquatic plant stands remain. They include *Bidens laevis*, *Hydrocotyle ranunculoides, Polygonum* sp and *Lemna* (Schmit-Munmm, 1998; Moreno *et al.*, 2005). These plants notably are able to withstand the harsh conditions of raw sewage-rich waters that are uncontrollably flowing from the nearby neighborhoods and from local supply centers such as the Central de Abastos Corabastos. This supply center is one of the main fruit, meat, vegetable and related goods receiver and distributor within Bogotá, which produces tons of solid wastes and sewage that end up in the wetland.

Aquatic and land vegetation face threats from illegal dumping of rubble and solid wastes, such as garbage that can pile up in different sectors. Invasive plant species such as the invasive grass *Pennisetum clandestinum* has taken advantage on the piled rubble and is often observe growing and filling virtually all spaces. Although a few exotic tree species, such as *Acacia*, *Eucalyptus*, *Fraxinus* and *Sambucus* have been reported around the wetland, today very few trees stand alone as a result of logging and a careless community attitude. The common mouse (*Mus musculus*) and rats (*Ratus ratus*) abound, but few species of birds remain, among them the common moorhen *Gallinula chloropus,* the yellow-hooded black bird *Agelaius icterocephalus,* the rufous-collared sparrow *Zonotrichia capensis* an introduced species common throughout the city, the eared dove *Zenaida auriculata* an also common bird that is not unique to the wetland (EAAB and CIC, 2000). This information may be unreliable, however, because of the misidentifications. Local people have been known to hunt birds like the yellow-hooded black bird, which is sold as a local pet. People normally trap these birds by sequestering the female that signs and attracts the male counterpart which has the distinctive yellow-colored chest. Traps with bait are another way to catch birds. Reptiles, amphibians and fish have not been studied and might be completely absent due to the disturbing environmental issues.

Physicochemical Aspects

There is no reliable information on physicochemical conditions in this wetland. The local community shows no interest and continues to ' connect' illegal sewage inflow tubes into the wetland to aggravate the wetland's current state. Bogota's Water Company has not measuredthe water chemistry.

Environmental Problems

Sewage

Raw sewage inflow from the nearby neighborhoods is the main source of water pollution. The local supply center known as Central de Abastos Corabastos, has been responsible for liquid discharges from leaks on its outer wall that is adjacent to the wetland. The supply center has a long history of sewage discharge and fines have been issued by Bogotá's office. For example in December 2009, the supply center had to pay a *$137,144,400 pesos (US$763,291)* fine for installing illegal raw sewage connections that flowed to La Vaca wetland.

Roads

The Cali highway, and a number or minor avenues and roads (Carrera 88) from Bogotá pass near this wetland, but their impacts have not been measured. Considerable amounts of runoff can be a source of pollutants flowing into the wetland.

Landfilling

Landfills have been made extensively in the area, thus reducing the wetland area. Landfills are used mainly for housing. They have been built around the wetland without any land use plan exerted by the government. Of the original wetland area, 95% of the total has been invaded by un-planned housing.

Restoration Efforts

Laws, Action Plans and Conservation

Colombia's regulations declared La Vaca wetland as an ecological district park according to Decree 190 of 2004 (IDEADE-EAAB, 2009). Bogota's Water Company and Bogotá's office have generated two independent management plans for La Vaca wetland in 2009 and 2010 respectively. The General Management Plan for La Vaca wetland was proposed by Bogotá's Water Company and the environmental rural institute (Instituto de Estudios Ambientales para el Desarrollo- IDEADE) of Pontificia Universidad Javeriana (IDEADE-EAAB, 2009). The plan included a diagnosis of environmental issues that the La Vaca wetland faced. The plan defined two management zones, one for terrestrial and the other for aquaticenvironments. Specific zones include a:

- Water management and improvement zone
- Aquatic and terrestrial recuperation zone
- Animal protection and preservation zone
- Recreation and education zone

The plan involves remediation, restoration and conservation. The plan's proposed strategies are: A) participative and applied research, B) education, participation and communication, C) remediation, protection and sustainable management and D) inter-institutional aid. Although, this plan can be classified as robust, few actions have been carried-out. Here history repeats itself, plans are proposed but action lacks.

Bogotá's city office, proposed another management plan for la Vaca Wetland (SDA, 2010b). The plan has the same overall structure described for other wetlands (Se section above in Santa María del Lago and Córdoba).

This general structure focused on the same points for example, always divides the wetland into several zones: restoration, remediation, conservation, etc, but not all wetlands are the same and thus, particular criteria should be generated accordingly. The current outcome of La Vaca' s plan is unperformed actions.

EL BURRO WETLAND

General Information

El Burro wetland is found to the southeast of Bogotá. This wetland borders paddocks on its southwest side, and neighborhoods, such as Urbanización Arboral Monterrey, El Castillo, and Villa María to the east. To the west the wetland borders the Tintalia neighborhood and adjacent paddocks. The wetland has been separated into two portions by the Cali Highway. Here, a levee was built to the northeast to block flooding. Some areas such as the one near America's Avenue have been completely urbanized as a result of landfills and the only wetland remains is a small channel that collects sewage and runoff water. This portion of the wetland currently has neighborhoods derived from landfill including Bosques de los Ángeles, Parque de los Ángeles and Los Ángeles. Most urbanizations date back to the early 90s when landfills began over the area. Bogotá's Water Company built a water system to link the two wetland portions and has also built local sewage systems that act as receivers from neighborhoods.

It has been estimated to cover 31 Ha (Figure 15) (Moreno *et al.*, 2005; Pinilla, 2010). El Burro has a hydric flow system limited by roads, avenues and highways especially during the dry season (Van der Hammen *et al.*, 2008). The wetland was once part of the Fucha River basin that has connections with El Tintal Lake and the Bogotá River basin, but due to landfills and roads constructed, El Burro now has few connections with its original basin.

Biodiversity

The only sources of information about the biodiversity within El Burro wetland are reports by Bogota's Water Company, these are inconsistent due to inappropriate sampling and species identification issues.

Terrestrial plants that are found scattered along the wetland include clustered dock *Rumex conglomeratus*, a herbaceous species native to northern Africa that is usually found near water banks and in dry areas. Trees, such as the castor oil *Ricinus communis* is native to the Middle-East and northern Africa, *Acacia* sp, an introduced species that can reach up to 30 m in height, and the fire thorn *Pyracantha cocconea* that has red berry-like fruits and is native to southern Europe extending to the Caucasus Mountains in western Asia are all exotic to the wetland. Much of the original vegetation has been replaced mainly with *Eucalyptus*. Few native species remain, for example the sauco, *Sambucus peruviana* characterized by white inflorescences and growing up to 15-20 m high (EAAB and CIC, 2000). Aquatic plants include those commonly found in the Bogotá wetlands, such as the floating plant *Lemna minor*, *Polygonum hydropyperoides*, the yellow-flowered *Bidens laevis,* the giant bulrush *Scirpus californicus* that forms dense matts, the common rush, *Juncus effusus*, and *Thypha* sp, (Schmit-Mumm, 1998, EAAB and CIC, 2000).

There are reportedly 32 bird species inthe el Burro wetland (EAAB and CIC 2000), but this information needs to be confirmed. To date, there are no reliable bird studies for this wetland. Moreover, there have been no studies on populations of fish, amphibians or reptiles. This wetland has been poorly monitored and is currently in a critical state.

Environmental Problems

El Burro faces multiple environmental problems such as sewage, solid residues, land filling, roads crossing the wetland many of which have insufficient information to provide a clear picture to the reader. Therefore, we will focus on those problems with more information.

Sewage

Sewage is one of the main problems of this wetland, but little information on sources and types of contamination are available. It has been known that from 1993 major raw sewage inputs have been flowing toward the wetland from nearby neighborhoods. Bogotá's office and Bogota's Water Company have not done much to remediate sewage inflow into this wetland.

Landfilling

Landfilling has been the main mechanism that has reduced the original wetland size. Landfills start by piles of debris and construction remains piled up in scattered areas within the wetland that after time are shaped into dry land for contractors to illegally use. Currently, there are no reliable estimates of the original wetland size lost to landfilling, and the invasive process is a reminder of the lack of interest and corruption among Bogota's governmental organizations.

Laws, Action Plans and Conservation

The SDA recently implemented a management plan for the El Burro wetland (SDA 2010c). The plan structure is nearly the same as for the other wetlands. The actions derived from this wetland are lacking and again, the government has repeatedly omitted actions that are within its jurisdiction and duty.

CAPELLANIA WETLAND

General Information

The Capellania wetland is part of the Fucha River basin, which flows into the Bogotá River. This wetland has been disconnected from its original basin and its hydric flow system is limited by roads, avenues or highways especially during the dry season (Van der Hammen *et al.*, 2008). The wetland´s outflow occurs through a channel that flows into Fontibón channel. The wetland was fractioned in two sections (referred as lower and upper sections) by the passage of Luis Cralos Galán Avenue built in 1995. The southern basin of this wetland has been densely populated (Moreno *et al.*, 2005). The wetland is immersed within the southwest of the city within the Fontibón locality. In the past, it covered 37 Ha currently, there are 18 Ha (Figure 16) (DAMA, 2006; Pinilla 2010).

Biodiversity

Similarly to other wetlands within the district of Bogotá, the Capellanía wetland has been severely degraded. Few animal species remain (Moreno *et al.*, 2005). For example, the thickhead ground snake *Atractus crassicaudatus*, has been found dead on the nearby roads as a consequence of automobile passage. Due to the harsh conditions most bird species are absent from this wetland (Moreno *et al.*, 2005).

Few aquatic plants remain. These include aquatic plants that are common throughout the city's wetlands, for example, the giant bulrush *Scirpus californicus*, the smooth beggarticks *Bidens laevis*, broadleaf cattail *Thypha latifolia*, the clustered dock *Rumex conglomeratus,* and the swamp smartweed *Polygonum hydropiperoides* (Schmidt-Mumm, 1998). The invasive grass *Pennisetum clandestinum* threatens most aquatic plants during the dry season when water level is low and this grass is able to colonize extensive areas. Some areas, such as the one near Occiedente Avenue have been used for cattle grazing. Tree species are scarce and the few occurring, such as *Eucalyptus* are mainly introduced (Moreno *et al.*, 2005). The willow *Salix humboldtiana* is among the few native tree species that remain. Trees are often found as solitary individuals scattered throughout the wetland.

Environmental Problems

Sewage

Waste water that enters into the wetland on its upper portion from nearby industries is rich in heavy metals. Raw sewage is also discharged directly into the system. Capellanía also faces great sedimentation problems (Moreno *et al.*, 2005).

Land Filling

Landfill has been the main mechanism for reducing the wetland area. Landfills made on the late 20th century contributed to the establishment of local factories and industrial complexes. Zoning for industry has been a major reason for heavy metal contamination of this wetland.

Laws, Action Plans and Conservation

Capellanía, considered one of the smallest wetlands within Bogotá is threatened to the point of disappearance. This has been caused by an endless history of environmental degradation by raw sewage inputs, industrial-related inflow, solid disposal, rubble, landfills and biodiversity loss. The first approaches to restore its original conditions were proposed within an environmental management plan written by Ecology and Environment Inc,

Hidromecánicas LTDA and the Bogotás office in 1996. But, due to inconsistencies this plan was not approved (Moreno *et al.*, 2005). Later on, Bogotá's city office proposed a new management plan for the Capellanía wetland in 2010 (SDA, 2010d). Due to similar structure to that of other wetlands, the plan will not be discussed herein (See plan structure for Santa María del Lago and Córdoba). This plans actions were never enacted.

JABOQUE WETLAND

General Information

During pre-colonial times the Jaboque wetland was a sacred place for the native Muiscas, inhabitants of la Savanna of Bogotá. Historically, in the early 20th century, the Jaboque wetland was nested upon an agricultural and cattle pasture landscape. The first impacts to this ecosystem were made in 1948 and 1950 by the El Dorado Airport and El Dorado Highway respectively. These works affected part of its original area by reducing and affecting the hydric flow (DAMA, 2000).

During the mid-50s the area around the wetland was mainly agricultural land. The agricultural land began is transformation during the 60s by scattered constructions, mainly homes particularly in the southern part. During the late 60s and 70s the areas adjacent to the wetland were used-up by nearby neighborhoods, namely private communities, such as Villa de Granada, La Perla and invasive neighborhoods, such as San José Obrero. During the 80s massive invasions occurred, landfills were extensively made, raw sewage, rubble and solid disposal occupied areas within and nearby the wetland. The wetlands hydric system was changed by levees constructed over some of its margins. Over the 90s, the wetland had neighborhoods and urbanizations along, and in some cases within its margins (DAMA, 2000). Currently, this wetland faces massive problems.

Satellite imagery shows that Jaboque' s total wetland area is 108.39 Ha. (Figure 17) (Rosselli and Stiles, 2012). This wetland is found to the west of Bogotá near the El Dorado Airport and immersed within the local neighborhoods of Engativa, Bolivia, Villa del Mar, Villa Amalia, Villas de Granada and Alamos Norte.

The Jaboque wetland receives water from the Bogotá River near La Florida Park on the eastern zone of the wetland. The wetland also intercepts the floods from the Bogotá River during high waters.

Biodiversity

The area near the wetland has few trees or vegetation due the long history of landfills and human invasion. The land around the wetland is composed of neighborhoods, which have left very little space for trees to stand. Most of the few remaining trees include the non-native species, *Acacia* and *Eucalyptus*. In contrast to the surrounding land, the aquatic plant community is well structured and is currently home to 14 species. Pinilla (2010) estimated the abundance of aquatic plants within this wetland using frequency of occurrence (FO) (See the biodiversity section above in Santa María del Lago wetland for further clarification of FO). Likewise, Hernández and Rangel-Ch (2009) identified the 14 plant communities in different environments at Jaboque such as vegetation found near the water's edge was dominated by California bulrush *Schoenoplectus californicus*, common rush *Juncus effusus* and the broadleaf cattail *Typha latifolia*. In contrast, Pinilla (2010) found low abundances for these species (California bulrush FO= 1.5, common rush FO= 6 and broadleaf cattail FO= 0). In marsh habitats, dominant species include the stream bank rabbitsfoot grass, *Polypogon elongatum*, the clustered dock, *Rumex conglomeratus*, and the cortadera, *Carex luridiformis*, that belongs to the Cyperaceae family (Hernández and Rangel-Ch, 2009). The stream bank rabbitsfoot grass and the cortadera were not found in the wetland, and the clustered dockhad a low relative abundance with FO= 12 (Pinilla, 2010). *Solanum americanum* (FO= 5.1), swamp smartweed, *Polygonum hydropiperoides* (FO= 1.5), flatsedge *Cyperus rufus* (FO= 7,5), the Madagascar ragwort *Senecio madagascariensis* (FO= 4.5), the one row yellow cress *Nasturtium officinale* (FO= 13.75) were present but not abundant (Pinilla, 2010). Terrestrial-aquatic transition zones are typically abundant with the smooth beggarticks *Bidens laevis*, the floating primrose-willow, *Ludwigia peploides*, the floating marsh pennywort, *Hydrocotyle ranunculoides*, and the water pepper, *Polygonum punctatum* (Hernández and Rangel-Ch, 2009). Pinilla (2010) found that the smooth beggarticks (FO= 72.62), the floating marsh pennywort (FO=77.87), were among the most abundant plants. The water pepper and the floating primrose-willow were absent from this wetland. Vegetation within the free water zones included the floating plants, water hyacinth, *Eichhornia crassipes*, an exotic species, the West Indian sponge plant, *Limnobium laevigatum*, and the gibbous duckweed *Lemna gibba* (Hernández and Rangel-Ch, 2009). According to Pinilla (2010), the floating plants, water hyacinth (FO= 41.12) and the water fern (FO= 24.5) were moderately abundant, whereas the West Indian sponge plant and the gibbous duckweed were absent. The species composition differences among these two authors could be due to species misidentification.

Numerous terrestrial arthropods were found on aquatic plants such as the common rush, *Juncus effusus*, water pepper, *Polygonum punctatum*, California bulrush, *Schoenoplectus californicus*, broadleaf cattail, *Typha latifolia*, and smooth beggarticks, *Bidens laevis* (Sánchez-N and Amát-García, 2005). The diversity of arthropods is considered high at Jaboque. This is evidenced by a total of 212 morpho-species of arthropods found on the wetland plants. The most abundant arthropod order was Diptera (62%), followed by Hymenoptera (17%), Arachnids (7%), Coleoptera (6%), Heteroptera (6%), Lepidoptera (1%), and Odonata (1%) (Sánchez-N and Amát-García, 2005). Phytophagous (plant eating) insects had the greatest richness and abundance among aquatic plants except for the California bulrush and the common rush (Sánchez-N and Amát-García, 2005). Nectar and pollen feeders had intermediate to low levels of richness and abundance (Sánchez-N and Amát-García, 2005).

Based on seed bank germination from patches within the Jaboque wetland, aquatic plants were classified as being highly vulnerable, vulnerable and not vulnerable (Montenegro-S *et al.*, 2006). This investigation used patches of aquatic plants. From each patch a total of six soil samples were taken to determine seed bank germination under greenhouse conditions. After the selected soil samples were transferred to the greenhouse, it was determined if the seeds were able to germinate over time and if they formed a seed bank. Highly vulnerable species were poorly represented in the plant community and were not present in the seed bank. These included species, such as *Carex luridiformis* and *Ludwigia peruviana*. Vulnerable species with low frequency and germination within the plant community included *Cyperus rufus*. Non-vulnerable species with high germination rates included *Ludwigia peploides, Typha latifolia, Juncus effusus, Polygonum* sp. and *Bidens laevis* (Montenegro-S *et al.*, 2006).

Phytoplankton is represented by diatoms, dinophytes, euglenoids, chlorophytes, and cyanobacteria (Pinilla, 2010). Among these the most predominant group in Jaboque was the chlorophyte community composed of *Oocystis* sp. and *Scenedesmus* sp. The occurrence of the chlorophyte might explain the high oxygen content (Pinilla, 2010).

Invertebrates, such as dipterans, are an abundant group in the Jaboque wetland during the rainy season (about 4,000 organisms m^2) and in the dry season (about 3,000 organisms m^2). Other invertebrate organisms included oligochaetes (worms), gordioids and lichees (Pinilla, 2010).

There have been 28 bird species reported in the Joboque wetland (Rosselli and Stiles, 2012). As previously mentioned (see section about Santa María del Lago) these birds have been classified into different guilds according to Rosselli and Stiles (2012). For example, the black-bellied whistling duck, *Dendrocygna autumnalis*, blue-winged teal, *Anas discors*, amigratory species that nests in North America and flies to South America during winter (Asociación Bogotana de Ornitología, 2000), the lesser scaup, *Aythya affinis*, another migratory species from North America that overwinters in South America (Hilthy and Brown, 1986), and *Oxyura jamaicensis*, are all ducks that belong to the surface swimmers guild that feed on invertebrates. The zambullidor *Podilymbus podiceps* belongs to the benthic diver's guild. The little blue heron, *Egretta caerulea*, cattle egret, *Bubulcus ibis*, striated heron, *Butorides striata*, the great egret, *Ardea alba*, black-crowned night heron, *Nycticorax nycticorax*, are all long-legged, egret-like birds that belong to the waders-waiters guild that feeds on invertebrates. The bare-faced ibis, *Phimosus infuscatus*, belongs to the guild of waders that probe in mud for invertebrates, and the fish hawk, *Pandion haliaetus*, is part of the flight feeders that plunge for fish guild. The Bogotá rail, *Rallus semiplumbeus*, the migratory bird, *Porzana carolina*, purple gallinule, *Porphyrio martinica*, and the common moorhen, *Gallinula chloropus,* all belong to the guild of walkers on aquatic vegetation that consume invertebrates and plants. The spot-flanked gallinule, *Gallinula melanops*, and American coot, *Fulica Americana,* are part of the surface swimmers guild.

The southern lapwing, *Vanellus chilensis*, which belongs to the shoreline walkers that consume invertebrates in mud or bare soil guild and the noble snipe, *Gallinago nobilis*, are in the waders that probe in mud guild. The lesser yellowlegs, *Tringa flavipes*, grand chevalier, *Tringa melanoleuca*, solitary sandpiper, *Tringa solitaria*, and spotted sandpiper, *Actitis macularius,* are all migratory species that are part of the shore walkers guild. The subtropical doradito, *Pseudocolopteryx acutipennis*, and the grassland yellow-finch, *Sicalis luteola*, belong to the guild of perchers on emergent vegetation that glean for insects. The migratory northern water thrush, *Seiurus noveboracensis*, and yellow-hooded blackbird, *Chrysomus icterocephalus*, which is a subspecies endemic to Colombia (Rosselli and Stiles, 2012).

Mammal species occupying the Jaboque wetland include the endangered mouse, Akodon (*Microxus bogotensis*) and the common guinea pig *Cavia porcellus*. Reptiles include the snakes, *Atractus crasicaudatus* and *Liophis epinephelus bimaculatus*. Amphibians, the endemic frog, *Hyla labialis* (EAAB and CIC, 2000).

Physicochemical Aspects

Pinilla (2010) reported various physicochemical variables from the Jaboque wetland in both rainy and dry seasons. Results were compared to some of Bogotás wetlands including Santa María de Lago, Tibaníca, Jaboque, Juan Amarillo and Guaymaral.

A small, rural wetland named Meridor (9.6 Ha) from the municipality of Tenjo on the Bogotá Savanna was analyzed as a reference ecosystem due its low degree of human intervention (Pinilla, 2010). During the rainy season, the water temperature in the Jaboque wetland was 20.8°C and in the dryseason was 22.5 °C. These temperatures are the highest ones recorded from all the wetlands.

The possible reason is unknown, but it is possible that industrial water inflow could be at a higher temperature. Total dissolved solids of 81 mg/L during the rainy and 117.8 mg/L during the dry seasons and conductivity at 153.2 μS – wet and 168.8 μS –dry were among the lowest values for the studied wetlands. The less intervened wetland Meridor, had much higher values of dissolved solids being 428 mg/L and conductivity 855 μS in the dry season. Dissolved oxygen was 14.40 mg/L in the rainy and 8.57 mg/L in the dry seasons, a considerably high value among the wetlands and excellent for aquatic living organisms. Water pH in Jaboque was only recorded during the dry season. It was 8.82, a relatively high level, which could mean waters rich in detergents are entering the wetland because these have a tendency to turn water alkaline.

Chloride levels at 22 mg/L during the rainy and 16 mg/L during the dry seasons were low to medium values. It should be mentioned that chloride concentrations above 200 mg/L are detrimental to aquatic life. Water hardness, at 80 mg/L ($CaCO_3$) during the rainy and 62 mg/L ($CaCO_3$) during the dry seasons, was intermediate in level compared with that of the other wetlands. Secchi disk transparency was 0.92 min the rainy and 0.74 m in the dry seasons.

These were among the highest values despite high levels of suspended solids (47.27 mg/L rainy and 45.56 mg/L dry seasons) when compared to the other wetlands. Nitrogen containing compounds (0.1 mg/L in rainy and 0.3 mg/L nitrate in dry seasons; 0.006 mg/L in rainy and 0.3 mg/L nitrites in dry seasons; 0.7 mg/L in rainy and 0.7 mg/L ammonium in dry seasons) were relatively low.

One should expect high nitrogen values due to high raw sewage input from the adjacent neighborhoods. Othophosphates at 1.4 mg/mL in rainy and

0.5 mg/mL in dry seasons was a low to medium value. Phosphate is essential for algal and plant communities as it is an element required in enzymatic reactions. The BOD had intermediate values of 24 mg/L (O_2) during rainy and 16 mg/L (O_2) during dry seasons, meaning that bacterial communities might be actively utilizing oxygen.

Environmental Problems

Sewage

Industrial and residential sewage inputs are the primary environmental concerns. Mainly, the southeastern zone receives inputs from sewage and rain-collecting channels (Arcos-Pulido and Gómes, 2006).

The northwestern portion of the wetland is considered less urbanized but has been impacted by cow and horse grazing, agricultural irrigation systems, and by the introduction of the grass, *Pennisetum clandestinum*, that has invaded waterways (Arcos-Pulido and Gómes, 2006).

Eutriophication

According to the limnological conditions index (LICOI) values analyzed by Pinilla (2010) at Jaboque, it is one of the most conserved ecosystems (LICOI higher than 17.5%) (Pinilla, 2010). Furthermore, Jaboque exhibited acceptable limnological conditions, which could indicate that the improvement efforts in sewage management in this ecosystem made by Bogotá's city office and Bogota's Water Company have been beneficial (Pinilla, 2010).

Periphytic algae that attach themselves to any surface have been used as biological indicators in Jaboque wetland (Arcos-Pulido and Gómes, 2006). Sixty-eight species of algae belonging to Euglenophyta (green algae), Chlorophyta (green algae), Cyanophyta (blue-green algae), Bacillariophyta (diatoms), Chrysophyta (Golden algae) and Cryptophyta (Cryptomonads) were studied at Jaboque. The most abundant group was diatoms (32.83%) followed by Chlorophites / Euglenophytes (27.37%), Cyanophytes (10.45%) and Chrisophytes (5.96%) (Arcos-Pulido and Gómes, 2006). The diversity, composition and diversity of periphytic microalgae indicate a high eutrophication process (Arcos-Pulido and Gómes, 2006).

Restoration Efforts

Bogotá's city office made restoration efforts by planting trees in 1995 such as the Andean oak, *Quercus humboltii*, a native tree considered an important timber species that can reach up to 40 m in height and *Croton bogotense* that can reach up to 20 m in height (Moreno *et al.*, 2005). To date there are no available accounts of the fate of these planted trees.

Laws, Action Plans and Conservation

Currently, there are no management plans by the government. This clearly shows little governmental interest. But, the first individual initiative to elaborate a proposal was made by Ospina (2007). This preliminary plan describes environmental policies, pollution monitoring by Bogotá's Water Company, and seeks to propose a legal management proposal for Jaboque. This plan analyses the water quality parameters measured by Bogotá's Water Company and determines the wetland´s pollution state. Furthermore, the plan establishes more actions such as environmental-related surveys applied to the local community (Ospina, 2007). The outcome of this first plan is that the local community has little or no interest and do not seem to take part in any actions toward Jaboque´s management or conservation.

TECHO WETLAND

General Information

The Techo wetland was originally known as Techo' s ranch (Hacienda Techo) owned by the Jesuits and sold to "private hands" in 1978. This wetland was part of the large Tintal Lake that was divided into smaller wetlands as a result of human intervention and establishment. This history of human intervention is particularly evidenced by a reduction in wetland size. This wetland covered 23 Ha in 1956 and has been reduced to the current 3 Ha, which are rapidly disappearing (Figure 18) (Moreno *et al.*, 2005; Pinilla, 2010). This means, that over the past 40 years or so, landfill and illegal construction have been made without any governmental intervention or planning. Illegal neighborhoods include Lagos de Castilla built in the 90s.

The Techo wetland is part of the Kennedy locality to the southeast section of Bogotá. It is fed by rain and flows into Bogota's river flood plain (Muñoz, 2006). During the 70s and 80s, Techo was isolated from the original Bogotá River drainage system and currently has no drainage exit (Muñoz, 2006). Among social problems, poverty threatens the wetlands as people invade, and try to get an opportunity to build homes.

Biodiversity

Techo is one the most polluted wetlands in Bogotá. The un-planed housing that has reduced the wildlife habitat and general environmental conditions has aggravated its ecological and environmental problems. The wetland has been isolated from its original drainage causing a non-connectedness and possibly an interruption in species movement. Moreover, cattle pasture has increased the impact into the more terrestrial parts of this ecosystem (PUJ-EAAB, 2007).

The terrestrial ecosystem around this wetland has been colonized by the invasive grass, *Pennicetum clandestinum*. There are also exotic trees like *Acacia,* fire thorn, *Pyracantha coccinea,* which grows up to 4.5 m in height and the castor oil tree, *Ricinus communis,* distributed randomly within the terrestrial zone around the wetland. Native tree species include the sauco, *Sambucus peruviana*, the guayacán de manizales, *Lafoensia speciosa*, a tree that can reach up to 30 m in height and can be used for timber, the alder, *Alnus acuminata*, and the locally known sauce *Salix humboldtiana*, that can reach up to 30 m in height and is found living partially submerged within the water. The arrayán, *Myrcianthes leucoxyla*, produces edible fruits for birds, and the carbonero rojo, *Calliandra carbonaria* has distinctive red flowers and can reach up to 6 m in height. Finally, the chicalá, *Tecoma stans*, is a yellow-flowered tree that belongs to the Bignoniaceae family (PUJ-EAAB, 2007).

Aquatic plant communities within the wetland have also been characterized. These can be seen covering most of the wetland. Among these, the floating marsh pennywort, *Hydrecotyle ranunculoides*, covers most of the wetland´s surface. The commonduck weed, *Lemna minor*, forms carpet like aggregations on the water surface. This plant has an overall size of 1 cm in diameter. The West Indian sponge plant, *Limnobium laevigatum*, is a floating plant 2 cm in diameter and often forms dense aggregations. The broadleaf cattail, *Typha latifolia,* that stands up to 2.5 m in height, common rush, *Juncus effusus,* that can reach up to 1.5 m in height, floating primrose-willow, and

Ludwigia peploides, that can reach up to 50 cm in height (Moreno *et al.*, 2005; PUJ-EAAB, 2007). These plants have adapted well to the human-caused intervention for more than 40 years.

Algae that occurs within the wetland include Bacillariophyta, Charophyta, Chlorophyta, Euglenophyta and Pyrrophyta (PUJ-EAAB, 2007). However, identification to species level needs revision. Zooplankton and invertebrates have been poorly studied.

Few investigations have focused on the fauna of the Techo wetland. Most literature on the topic is basically internal reports from the government and private organizations. The wetland is home to 28 unconfirmed bird species (EAAB and CIC, 2000). Due to their unconfirmed status these will not be discussed. Other preliminary reports include about 38 bird species (PUJ-EAAB, 2007). However, this information is limited in scope because observations were conducted only in two days in April, 2006. Clearly, more in depth and consistent research is required. Reptiles are represented by the endemic snake *Atractus crasicaudatus* that is often found along the other wetlands of the city. Amphibians include the green frog, *Hyla labialis*, which is also common among the wetlands (EAAB and CIC, 2000).

Physicochemical Aspects

Poor limnological conditions have been found at the Techo wetland. This wetland has been impacted severally by sewage and industrial inputs. Physicochemical conditions have been determined for different zones by governmental and the private sector. However, this information is inconsistent because of poor replication, making it difficult to draw conclusions. See PUJ-EAAB (2007) environmental plan for further details.

Environmental Problems

Sewage

The Castilla Real urban project was built on in 1973. This housing project placed all derived sewage into Techo´s wetland (PUJ-EAAB, 2007). The urbanized process within the wetlands clearly shows an increased tendency in the number of channels and tubes that collect raw sewage and pour it into the wetland. By the early 1990s Bogota´s Water Company constructed a series of channels on the southeastern portion of the wetland that caused a decline in the local *Juncus effusus* population (PUJ-EAAB, 2007). Apart from this, the wetland receives agricultureal animal-residues, and industrial sewage inputs.

Fragmentation

Habitat fragmentation has been claimed to be an important aspect for the reduced number of animal species observed (PUJ-EAAB, 2007); however, caution must be taken because no reliable population estimates have ever been made.

Landfilling

Historically, the government has not taken protective measures on landfilling (Moreno *et al.*, 2005). Landfilling was first established in 1976 as a mechanism to establish human housing locally for people with few resources (Muñoz, 2006). Flooding of the northwestern margin occurred as a consequence of landfilling (Muñoz, 2006). Illegal occupation took place in 1989 and is currently one of the main issues that the wetland faces. The main unplanned, illegal or "pirate" community is Lagos de Castilla, a twenty-three block area (PUJ-EAAB, 2007).

During the late 80s and early 90s, La Colina Real Estate Agency, a legal group, sold illegal landfilled property to poor people (PUJ-EAAB, 2007). After this, the land passed to other owners and urban development within the area has been stopped.

Introduced landfilling material has caused the wetland to divide into three zones. The first zone, located on the southeast portion of the wetland which been greatly urbanized. The second zone, located on the northwestern portion has been poorly urbanized. The third centermost zone is the largest and has the best ecological conditions (PUJ-EAAB, 2007).

Roads

The Techo wetland has been severally impacted by roads. Since 1954 it has been surrounded by nearby neighborhood roads. Recently, Cali Avenue on the northwest margin cuts part of the wetland (Muñoz, 2006). Un-paved roads cross sections of the wetland on the southern zone. These roads generate an impact that has not been measured.

Restoration Efforts

The Techo's wetland restoration potential is relatively low due to the greatly fragmented zones, landfill problems, and local biodiversity loss (Stiles and Rosselli, 2004). The government has paid little attention to the wetland. Currently, there is no political interest and no policies that specifically promote restoration efforts. This is largely due to local corruption and the lack of self-identity of politicians and citizens.

Laws, Action Plans and Conservation

Decree 457 of 2008, declared the "critical state" of Techo' s wetland and stressed the need for public space reinstitution. Bogotá's Water Company proposed an environmental evaluation for the wetland in 2000. This evaluation focused on 1) current environmental issues and 2) ecological potential. Within each of these categories, physical, biological, social and cultural parameters were evaluated. Physical parameters included flood control, sediment retention, microclimate and water quality management (PUJ-EAAB, 2007). Biological parameters included threatened or endemic species protection, migratory aquatic and terrestrial bird species habitat, plant and animal richness, hydrological richness, habitat richness, interactions with other ecosystems and top trophic level species (PUJ-EAAB, 2007). The social and cultural parameters comprise people's common use, research use, and landscape value (PUJ-EAAB, 2007). This evaluation was successfully conducted prior to defining the environmental plan for the Techo wetland.

Recently, Techo' s wetland environmental management plan was proposed by the rural environmental institute of Pontificia Universidad Javerina and Bogotá's Water Company in 2007 (PUJ-EAAB, 2007). This plan was constructed with the local community by applying surveys that reviewed biological, land use, social and environmental impacts faced by the wetland. The plan established working groups between the local community, Universidad Javeriana and Bogotá's Water Company. These groups defined possible solutions and community commitment with the environment. Working groups identified seven critical aspects 1) legal conflicts, 2) pollution, 3) communication and coordination within institutions, 4) education, 5) urbanization, 6) water issues and 7) ecology (PUJ-EAAB, 2007). Although the action plan proposes solutions, there are conflicts among the community, institutions and laws, mainly because of land use issues and conflicts of interest. The plan has defined conservation and protection areas and wetland priorities, but it only describes possible conservation areas and priorities, but does not propose local actions or a reference frame among the community, government and other participating institutions. Thus, due to lack of policy endurance, applicability and corruption, the plan has no effective outcomes. It is hoped that Techo´s wetland action plan is given the attention it deserves and that actions are put forth to preserve what was once a great wetland.

The local community has generated protective environmental measures. There has been an educational process where local leaders, school teachers, students and the general public have been informed about wetland protective measures. Since the establishment of Techo´s wetland administration by 2008, four main management areas have been adopted that include vigilance, ecological management, social issues and monitoring. Although, the government assigned a budget for management practices, the NGO, Fundación Alma, which is currently in charge of the wetland´s management, has made little progress to overcome the multiple environmental issues.

TIBANICA WETLAND

General Information

Tibanica was part of the large Tintal Lake. Due to Techo´s aerodrome construction in the 1930s and America' s Avenue built in 1946, Tintal Lake was fractioned into smaller water bodies that comprise wetlands such as Tibanica, La Vaca, El Burro, Techo and Timiza Lake. After this intervention, the area around Tibanica began its current development into neighborhoods within the Bosa locality, immersed on the southwestern portion of Bogotá. The Tibanica wetland covers a total area of 18.98 Ha. (Rosselli and Stiles, 2012) (Figure 19). The wetland is nested next to Manzanares and la Esperanza neighborhoods in the northern portion. The northeastern portion of the wetland borders the Júpiter, Barlovento, los Barrios, and Los Olivos 2º sector neighborhoods and nearby paddocks at El Junco, La Tingua and El Erial. On the southern portion, La Despensa, León XIII, Pablo VI, Juan Pablo I, La María, and Rincón de Santa Fe neighborhoods have been established by landfilling. The southeastern portion is covered by paddocks, agricultural and floricultural lands. A fallen bridge that divides it into two portions has cut the wetland. One portion is named Potrero Grande (Big paddock) and has about 26 Ha, and the other one is the Tibanica wetland. The Potrero Grande portion has been drained and is used as cattle pasture land. This land has been devoid of any wetland plant and animal life and has been colonized mainly by the exotic grass, *Pennisetum clandestinum*.

The wetland belongs to the Bogotá river basin. The Tibanica stream, its main Tributary, is no longer connected to the wetland, and a levee was constructed in 2009 by Bogotá's Water Company. This levee was built to isolate the wetland from Tibanica stream due to its severely polluted conditions mainly by domestic and industrial sewage inputs from the nearby Soacha municipality.

Biodiversity

The space around the wetland is nested within neighborhoods and paddocks. The landscape has been modified and the areas around the wetland are covered by exotic grass (*Pennisetum clandestinum*). Non-indigenous tree species like the alcaparro, *Senna viarum*, a tree that can reach up to 20 m in height and *Eucalyptus* are found around the wetland. The native tree species found along the wetlands bank is the willow, *Salix humboldtana* (Moreno *et al.*, 2005). Recent restoration action since 2009 by Bogota's Water Company was to plant a total of 1,500 individual trees on the terrestrial portion of the wetlandand to remove exotic water hyacinth. However, the current restoration actions have been discontinued due to lack of interest.

The water zone has aquatic plants, either submerged or floating that include 14 species (Pinilla, 2010). They include cluster dock, *Rumex conglomeratus,* an exotic species from northern Africa, *Pennisetum clandestinum*, and the clover, *Taraxacum* sp., which are mainly terrestrial plants that are often found within the wetland. The abundance, using frequency of occurrence, was determined for the California bulrush (*Schoenoplectus californicus*). It was the most abundant aquatic plant (FO= 60) forming dense patches within the wetland Other abundant plants are the common duckweed (*Lemna minor*) (FO=38.75), the gibbous duckweed (*Lemna gibba*) (FO= 36), and duckweed (*Spirodela intermedia*) (FO= 51) that are small floating plants. Another abundant plant was the broadleaf cattail (*Typha latifolia*) (FO= 50) that is common throughout the city wetlands. Less abundant aquatic plants (FO $\leq$ 8) that inhabit this wetland comprise smooth beggarticks, *Bidens laevis*, which is among the most abundant plant in other wetlands within Bogotá district, Least duckweed, *Lemna minuta*, the water fern *Azolla filiculoides*, and the floating marsh pennywort, *Hydrocotyle ranunculoides*, are rather abundant in other wetlands.

Phytoplankton is comprised of diatoms, dinophytes, euglenoids, chlorophytes, and cyanobacteria (Pinilla, 2010), but the most predominant group at Tibanica is cyanobacteria.

Dipteran was the most abundant invertebrate group at Tibanica during the rainy season (about 10,000 organisms m^{-2}) and the dry season (about 8,000 organisms m^{-2}) (Pinilla, 2010). Other invertebrate groups included gordioids, leeches, and bivalves (Pinilla, 2010). Macroinvertebrate numbers were low, possibly due to an excess of organic matter in sediment, which increases oxygen demand (BOD5) and worsens the benthic environment conditions for these organisms (Pinilla, 2010).

Tibanica is home to 23 bird species (Rosselli and Stiles, 2012). Among them are migratory species like ducks, *Anas discors*, *Aythya affinis*, the sora, *Porzana carolina*, the upland sandpiper, *Bartramia longicauda*, the pectoral sandpiper, *Calidris melanotos*, the grand chevalier, *Tringa flavipes*, greater yellowlegs, *Tringa melanoleuca*, solitary sandpiper, *Tringa solitaria,* and the spotted sandpiper, *Actitis macularius*. The wetland is also home to the endemic threatened wren, *Cistothorus apolinari,* which uses *Schoenoplectus californicus*, *Pennisetum clandestinum*, and *Rumex obtusifolius* for nesting (Rosselli and Stiles, 2012). Mammals include the guinea pig, *Cavia porcellus,*and the long-tailed weasel *Mustela frenata*. Reptiles include the snake, *Atractus crasicaudatus*. Amphibians are represented by the frog, *Hyla labialis* (EAAB and CIC,2000).

Physicochemycal Aspects

Pinilla (2010) measured physicochemical conditions in the Tibanica wetland during the rainy and dry seasons. The water temperature was 17.1°C during the rainy and 19.1°C during the dry season. These temperatures are around the average found in Bogotá wetlands. The total dissolved solids was 709 mg/L during the rainy and 819 mg/L during the dry season, which is the highest value among Jaboque, Juan Amarillo, Santa María del Lago and Guaymaral (Pinilla, 2010). These values could be indicative of high raw sewage input. Water conductivity was 1,210 µS during the rainy and 1,840 µS in the dry season, which are extremely high values and could indicate the presence of industrially derived ions. Dissolved oxygen at 0.2 mg/L in the rainy and 1.05 mg/L during the dry season are basically low levels that seem dangerous for typical aquatic life. The pH of 6.8 remained stable in the rainy and dry seasons. Chloride at 108 mg/L during the rainy and 74 mg/L during the dry season is relatively high when compared to other wetlands and could indicate industrial inputs from nearby factories. Water hardness was 91 mg/L ($CaCO_3$) during the rainy and 179 mg/L ($CaCO_3$) in the dry season. These are moderate levels for the studied wetlands. The secchi disk transparency was 1.27 mand 1.35 m, respectively. These values are remarkably high considering that this wetland has sewage inputs. Suspended solids levels were 9.16 mg/L during the rainy and 11.34 mg/L during the dry season, which are moderate values among the other wetlands. Nitrogen containing compounds, included nitrates at 0.01 mg/L, nitrites at 0.003 mg/L and ammonium at 1.2 mg/L during the rainy season. These are comparatively low values. Orthophosphate levels were 0.12 mg/ml in the rainy and 1.4 mg/ml during the dry seasons and are similar to the other wetlands. BOD levels were 29 mg/L (O_2) during the rainy and 8 mg/L (O_2) during the dry season. They are moderate values when compared among other Bogotá wetlands.

Environmental Problems

Environmental problems encompass illegal housing, solids disposal, rubble disposal, wood burning for making charcoal and land filling has occurred over the eastern portion of the wetland.

Eutrophication

Tibanica is one of the most deteriorated ecosystems according to the limnological conditions index (LICOI index lower than 4%) (Pinilla, 2010). Tibanica was the wetland that received the lowest score according to Pinilla, (2010) indicating that its ecological health is very precarious and is the most threatened wetland in Bogotá. The phytoplankton and aquatic invertebrate Biological index value (%BI) indicate that this wetland has regular limnological conditions, but diatom and aquatic plants (%BI) correspond to a system of poor limnological status (Pinilla, 2010).

Laws, Action Plans and Conservation

Decree 203 of 2003 declared that preventive measures should be taken at Tibanica. A management plan has been proposed by Bogotá's office in cooperation with the environmental institute IDEA of Universidad Nacional de Colombia in 2004. But this plan was not put forth to action due to lack responsibility by the government. Recently, The State Management Office (Corporacion Autonama Regional de Cundinamarca, CAR) established another management plan for the Tibanica wetland in 2010. The plan focuses on a general wetland overview of the ecology, wetland classification, conservation and management projects (CAR, 2010). According to the plan, the main environmental problems within Tibanica are landfills, sewage input, illegal solid residues disposal, vegetation loss, illegal occupation by human settlements, illegal housing permits (CAR, 2010). Management plans have been proposed several times, there is clearly a lack of continuity between administrations and to date the environmental management at Tibanica has produced no environmental changes nor actions.

GUAYMARAL-TORCA WETLAND

General Information

Guaymaral-Torca is divided into two wetlands, namely Guaymaral and Torca; however in this section, both wetlands will be treated as one if not stated otherwise. Located on the northern portion of the city, this wetland is part of the Usaquén and Suba localities. During the 1950s, this wetland was divided into two sections by the Northern Highway (Autopista Norte). The large section covers about 49 Ha and a smaller one of 22 Ha. Overall, the wetland covers 73 Ha (Figure 20) (Moreno *et al.*, 2005; Pinilla, 2010). But, recent satellite imagery has determined that it covers 31.89 Ha. (Rosselli and Stiles, 2012). The construction of 170 street (Calle 170) was also a major drawback on the wetland´s history because it further divided the larger portion. Along with the construction of roads and avenues, this wetland is immersed within Mirandela, Villa del Parado, Cedritos and San Cristobal neighborhoods that have been developed near to or within the wetland´s limits. The wetland is also located adjacent to cemetery lands (Jardines de Paz) along the North Highway. Much of the Guaymaral-Torca wetland has been lost to the sides of the North highway and currently has been transformed into paddock landscapes (Figures 21A-D and 22A-B). The wetland proper e.g. large open waters borders the Jardines de Paz cemetery to the south, with the Mazda dealership and National Reforestation Company to the east, and soccer fields to the north.

Inflow into the wetland is provided by two waterways, Torca channel and el Guaco stream. Drainage flows through the Guaymaral channel that flows into Bogota's River basin. This channel carries sediment rich and polluted waters. Remnants of this wetland are found along the North highway (Autopista Norte), these portions are located along the highway's sides and can be identified by the occurrence of aquatic plants (Figure 21A-D).

Biodiversity

The most common parts of this wetlands are the plant matts of terrestrial and aquatic plants that intertwine along the North Highway (Figure 22A-B). These matts are composed native trees like *Salix humboldtiana* that has adapted to live in this severely impacted environment. The native trees, *Alnus*, *Sambucus peruviana* and *Lafoesia,* can also be observed along these matts associated with aquatic plants. Exotic species include the widely spread *Ricinus* and *Acacia* among others. This environment is impacted by the constant passage of buses, truck and cars that use the North Highway. This highway is one of the main roads connecting Bogotá with the nearby villages of Zipaquirá and Chia and is a major road for traveling to other cities located on the eastern side of the Andean Cordillera. The terrestrial zones along the banks that extend beyond the cemetery and near the soccer fields have few tree species mainly the already mentioned *Salix humboldtiana, Alnus*, *Sambucus peruviana* and *Lafoesia*.

The water zone is home to 17 aquatic plant species (Pinilla, 2010). These plants form large aggregations and have been thoughtfully characterized by their abundance according to Pinilla (2010). All of these plants are common among the wetlands of Bogotá. The most abundant is the floating marsh pennywort (*Hydrocotyle ranunculoides* FO=87), which is a hardy aquatic plant known to be able to whithstand severely polluted conditions followed by flatsedge (*Cyperus rufus* FO=55.87) that is common near the bank forming dense aggregations. The exotic water hyacinth (*Eichornia crassipes* FO=51.87) was introduced from Africa. It is a floating plant that has covered the open free-water zone. Spikerush (*Eleocharis* sp. (FO= 27.37) can be found along the banks and is often used for nesting by bird species. Swamp smartweed *Polygonum hydropiperoides* (FO= 27.37) and the common rush (*Juncus effusus* FO= 25.62) are also commonly seen. Less abundant species (FO $\leq$ 16) include the yellow-flowered, *Bidens laevis*, the floating plant, *Spirodela intermedia*, the Madagascar ragwort, *Senecio madagascariensis*, broadleaf cattail, *Typha latifolia*, California bullrush, *Schoenoplectus californicus*, one row yellow cress, *Nasturtium officinale*, and American everlasting, *Gnaphalium americanum* (Pinilla, 2010). Although this wetland has a similar aquatic plant composition as that of other wetlands within the city, the plant abundances are different and this could be due to raw sewage input hotspots and places with lower sewage inflow. The grass, *Pennisetum clandestinum,* was not abundant (FO= 23) but it has been observed to take over the other plant species and reduce the abundance of aquatic species.

Rosselli and Stiles (2012), reported 7 bird species at Torca. Most of the species that inhabit these wetlands have been discussed in previous sections of this chapter. These include *Bubulcus ibis*, *Rallus semiplumbeus*, *Gallinula chloropus*, *Gallinago nobilis*, *Tringa solitaria*, *Sicalis luteola* and the endemic species to Colombia, *Chrysomus icterocephalus* (Rosselli and Stiles, 2012). This is considered a low diversity and it is associated with the proximity to the North Highway (Autopista Norte). Furthermore, Guaymaral is home to 14 bird species (Rosselli and Stiles, 2012). These include *Anas discors*, *Oxyura jamaicensis*, *Bubulcus ibis*, *Rallus semiplumbeus*, *Gallinula chloropus*, *Gallinula melanops*, *Fulica americana*, *Vanellus chilensis*, *Gallinago nobilis*, *Calidris melanotos*, *Tringa flavipes*, *Tringa solitaria*, *Serpophaga cinerea*a and *Chrysomus icterocephalus* (Rosselli and Stiles, 2012). The spot-flanked gallinule, *Gallinula melanops bogotensis* has been found to be more abundant near floating aquatic plant areas within the wetland (Osbahr and Gómez, 2011).

Vertebrates include mammal species such as the common guinea pig *Cavia porcellus* and white-eared opossum, *Didelphis albiventris,* which inhabits forests and open land up to 3,900 m.o.s.l. Reptile species include the snake, *Atractus crassicaudatus,* and the chameleon, *Phenacosaurus heterodermus,* which is an inhabitant of Bogotá's eastern hills. Amphibians include the endemic frog, *Hyla labialis* (EAAB and CIC, 2000; Asociación Bogotana de Ornitología ABO, 2000).

Aquatic invertebrates comprise mainly dipterans in both rainy (11,000 organisms m^{-2}) and dry seasons (11,000 organisms m^{-2}) (Pinilla, 2010). Other invertebrates included oligochaetes (worms), leeches, planarians and bivalves (Pinilla, 2010).

Phytoplankton includes diatoms, dinophytes, euglenoids, chlorophytes, and (Pinilla, 2010). The most dominant groups include euglenoids and cyanobacteria (Pinilla, 2010). These algae are indicators of poor water quality with high nutrient and organic matter levels (Reynolds, 2006).

Physicochemical Aspects

Pinilla (2010) recently measured various physicochemical aspects within Gaymaral-Torca Wetland during rainy (November 2007) and dry (February, 2008) conditions. Dissolved oxygen values ranged from 2.65 mg/L and 3.19 mg/L in the rainy and dry seasons, respectively. These values are relatively high considering that this wetland receives sewage inputs that should lower dissolved oxygen.

Suspended solids levels were 47.26 mg/L in the rainy and 796.75 mg/L in the dry season. These values are high and are notably a consequence of sewage discharge and high nitrate levels. Nitrogen containing compounds had intermediate levels (during the rainy season nitrate concentration was 0.01 mg/L, nitrites were 0.0002 mg/L, and ammonium was 0.95 mg/L) whereas dry season nitrates and nitrites were absent, and ammonium concentration was 1 mg/L.

Environmental Problems

Environmental issues faced by Guaymaral-Torca wetland have been recognized as severe over the last 50 years. These wetland systems are impacted by raw sewage, high frequency sound produced by passing vehicles, and extensive drying of the land by road construction.

Landfilling

The Torca wetland has been reduced by desiccation, namely landfilling since a project of Bogota's city office to eliminate its principal water input to avoid flooding of the highway has facilitated invasion by terrestrial plants (Rosselli and Stiles, 2012). Torca has also been reduced by landfilling for high class urbanizations. This is the case of San Simon, a private community complex that has devastated most of Torca.

Restoration Efforts

Guaymaral could be recovered relatively easily by improving connections to the creeks of nearby hills, which would bring clean water (Pinilla, 2010). Guaymaral is considered one of the most conserved ecosystems according to the index of limnological conditions (LICOI index value higher than 17.5%) (Pinilla, 2010).

Laws, Action Plans and Conservation

Bogota's water company proposed a management plan for Guaymaral-Torca in 1997. As has occurred with other wetlands, this plan produced no actions. The NGO Fundación Humedal de Torca Guaymaral focuses in restoration and conservation aspects. For more information see http://fundacionhumedaltorcaguaymaral.blogspot.com/2010/09/comercio-electronico.html

MEANDRO DEL SAY WETLAND

General Information

The Meandro del Say wetland has existed on ranches since colonial times. The landscape was agricultural with potatoes, corn, cereals and vegetables being cropped regularly from nearby lands. The landscape remained an agricultural land, in the 70s. (DAMA, 2000). This wetland was later isolated from its original drainage basin on the Bogotá River in 1986. Thus, Meandro del Say is a wetland without hydric flow, e.g. no water channels flow in or out. This caused the amount of water that enters the wetland to be reduced and to affect the local aquatic and bank population of plants. Landfills were made along Centenario Avenue during the 80s. The wetland suffered desiccation, and many hectares were lost as a consequence of low water inputs during the 90s. Industrieswere established along the Centenario Avenue, which deposited raw sewage into the wetland. Landfills were made in 1998, which removed much or the original wetland´s area.

Located on the southern part of Bogotá within the Fontibón locality, the Meandro del Say wetland covers 26.2 Ha (Figure 23) (DAMA, 2006). However, this number might be subject to reduction due to the current landfilling operations within the area. The wetland currently borders the industrial zone of Fontibón.

Biodiversity

The current status of this wetland unknown. The land around this wetland has been virtually devoid of trees except for a few *Eucalyptus* and *Acasia*, and most of the vegetation is composed of the invasive grass, *Pennisetum clandestinum*. Over the 90s the aquatic plants that were found in the wetland were *Scirpus californicus, Typha* sp, *Bidens laevis*, *Hydrocotile rannunculoides, Lemna* sp, *Polygonum seguetum* (Schmidt – Mumm, 1998). The status of these aquatic plants need to be evaluated, and it is hoped that is not too late before the whole wetland is converted into a landfill.

Meandro del Say is home to 22 unconfirmed bird species (EAAB and CIC, 2000). Among these are *Agelaius icterocephalus, Gallinango nobilis, Elanus caeruleus, Gallinula cloropus, Ardea alba, Dendroica fusca,* and *Contopus virens* (EAAB and CIC, 2000), but again, these are unconfirmed reports.

Environmental Problems

There is no reliable information or reports by governmental institutions on the current or past state of this wetland. This is one of the forgotten wetlands where massive problems like raw sewage inputs and landfills have been catastrophic, and no one has monitored the changes over time or taken a serious interest in initiating corrective action.

Laws, Action Plans and Conservation

The State's management office (Corporación Autónoma Regional de Cundinamarca- CAR) proposed a management plan for Meandro del Say in 2006 (CAR, 2006). The plan defined environmental strategies. It focused on applied research, community involvement based on environmental education, restoration, management and sustainable use, and inter-institutional management (CAR, 2006). The plan received no support from the local community resulting in no impact or action. The wetland is located in an economically poor region of Bogotá and is plagued by dramatic cases of violence and insecurity.

Conclusion

Wetlands in Colombia's Bogota are historically and geographically important features of the landscape. However, during the last half a century, Bogota's wetlands have been reduced massively from 50,000 ha to 800 ha. What remains is a highly fragmented network of mainly 12 wetlands. This reduction has been attributed to rapid urbanization in and around Bogotá. In spite of the reduction in area, the existing wetlands support a very rich biodiversity with mostly an indigenous flora and fauna. Eutrophication is the major threat to Bogota's wetlands today. Due to urbanization these ecosystems are enriched with ions, nutrients and organic wastes. Dumping of solid waste and sewage, landfilling for illegal housing, pesticide residues from nearby agricultural land, invasive species of plants and animals that compete with the existing flora and fauna, etc. increase the vulnerability of the biodiversity.

All of Bogotá's wetlands have a history of human intervention. They are all ecosystems critically threatened by landfills, solid waste disposal, raw sewage inputs, invasive species, rubble, roads, disconnection from original drainage basins, and a general lack of interest by the government and in most cases by the communities they serve. Despite these environmental problems, Bogota's wetlands still harbor a rich number of birds and aquatic plant species. This scenario is less optimistic for mammals and reptiles that have been completely removed from several wetlands. The plant and animal life have adapted to live within polluted environments, but this is only to a certain point until conditions are so drastically deteriorated that life in the ecosystem is no longer possible. Although, Bogota's city office and Bogotá's Water Company have implemented restoration and management plans for these wetlands on several occasions, there has been much controversy because actions are never enacted. This phenomenon is due to corruption and lack of interest on behalf of work officials and chiefs that these institutions represent. The local community of these wetlands has different views. Some, are supportive and have generated positive actions such as the one from Santa María del Lago, whereas other communities especially from El Burro, La Vaca and Meandro del Say are reluctant and have no interest in these wetlands. Colombia has a strong policy for wetland protection, but unfortunately these legislations are often obscured by politics, private interests, bureaucracy and corruption.

NGOs have worked together with the local community to solve wetland environmental issues, but some actions by local NGOs have been criticized as money and resources spent have been lost within fraudulent administrations. Other actions have yielded positive outcomes such as effective management of pollution and raw sewage.

All the wetlands mentioned in this chapter hold the status of district public parks and are protected by laws. Moreover, Bogotá is home to more wetlands that currently do not hold the status of district parks. This is the case of the el Burrito wetland nearby to el Burro wetland. Other areas such as the one along the North Highway (Autopista Norte) have been seriously degraded. Other, less known wetlands might have already been destroyed or are facing threatening situations. The city clubs have made wetlands part of their own private landscape. This is the case of El Club los Lagartos, Los Arrayanes, Country Club, Club Coquenzá, which have historically used the heritage that belongs to the public to be enjoyed by few.

Despite the many issues that the city's wetlands face, it is hoped that the biodiversity that is still part of them will be there for future generations. One way to do this is by conducting reliable investigations that encompass not only lists and descriptions of the wetland's inhabitants, but also interactions among environmental impacts and action-based remediation solutions. Much more work is required to make community commitment and education the basis for wetland conservation. Local community education and the education of future generations in schools is key to conservation of these precious wetlands. The government should also reflect upon its actions and better define non-corrupt policies and ways to counteract the current lack of operability. On the international level, there might be more possibilities as resources are often better administered and environmental plans and actions often have an outcome. Although Bogota's wetlands face great problems, there is hope that actions will be taken to stimulate the common citizen to preserve these historically magnificent places once teeming with life and creatures that have evolved upon the great Andean landscape. Moreover, these wetlands belong to humanity and, thus, should be places for observing and admiring the spectacle of living forms.

ACKNOWLEDGMENT

Thanks to L. Marien for her valuable support in this investigation

References

Álvarez-León, R. (2009). Notas sobre la avifauna del humedal Santa María del Lago, Cordillera Oriental de los Andes colombianos. *Revista Luna Azúl*, 28, 24-39.

Arcos-Pulido, M.P. and Gómez, A.C. (2006). Microalgas perifíticas como indicadoras del estado de las aguas de un humedal urbano: Jaboque, Bogotá., D.C. Colombia. *Nova*, 4, 60-79.

Asociación Bogotana de Ornitología (2000). Aves de la Sabana de Bogotá, guía de campo. Bogotá, Colombia: ABO, CAR.

Bejarano, P.A. and Bonilla, M.A. (2009). Dinámica espacio-temporal del humedal. Juan Amarillo entre 1950-2005. *Acta Biológica Colombiana*, 14, 87-106.

Bellinger, E.G. and Sigee, D.C. (2010). Freshwater algae: identification and use as bioindicators. UK, John Wiley and Sons.

Calvachi, B. (2003). La fauna de los humedales de Bogotá y la sabana. In Los humedales de Bogotá y la sabana Vol. 1. (Guarnizo, A and Calvachi, B. eds),pp. 109-140. Bogotá, Colombia: Acueducto de Bogotá,Conservación Internacional.

CAR. (1996). Resultados de análisis fisicoquímicos. Propuesta Plan de Manejo. Ambiental del Humedal Santa María del Lago, mayo 21. Bogotá, Colombia:Informe Técnico.

CAR (2006). Plan de manejo ambiental del humedal Meandro del Say. Corporación Autónoma Regional de Cundinamarca. Bogotá.

CAR (2010). Plan de manejo ambiental del Parque Ecológico Distrital Humedal Tibanica. CorporaciónAutónoma Regional de Cundinamarca. Bogotá. 24p.

Cerón, L. E. and Ramírez, E. (2011). Actividad microbiana en los suelos y sedimentos en el sietema Córdoba Juan Amarillo, Bogotá, D.C. *Revisita de la Academia Colombiana de Ciencias Exactas, Físicas y Naturales*, 35, 349-361.

DAMA, (2000). Historia de los humedales de Bogotá con énfasis en cinco de ellos. Bogotá, Colombia.

DAMA, (2006). Departamento Técnico Administrativo del Medio Ambiente. Informe anual de calidad del aire de Bogotá, 2005.

Dokulil, M.T. and Teubner, K. (2011). Eutrophication and climate change: Present situation and future scenarios. In Eutrophication: causes, consequences and control. (Ansari, A.A., Singh Gill, S., Lanza, G.R. and Rast, W. eds.), pp. 1-16. New York. Springer.

EAAB (1988). Resultados de análisis fisicoquímicos. Propuesta Plan de Manejo Ambiental del Humedal de Santa María del Lago, abril. Empresa de Acueducto y Alcantarillado de Bogotá. Bogotá, Colombia: Informe Técnico.

EAAB and CIC (2000). Los humedales de Bogotá y la Sabana. Bogotá, Colombia: Empresa de Acueducto y Alcantarilllado de Bogotá. Conservación Internacional Colombia.

Fjeldså, J. (1984). Three endangered South American grebes (Podiceps): case histories and the ethics of saving species by human intervention. *Annales Zoologici Fennici*, 21, 411-416.

Fjeldsa J. (1985). Origin, evolution, and status of the avifauna of Andean wetlands.In Neotropical Ornithology. (Buckley, P.A., Foster, M.S., Morto, E.S., Ridgely, R.S., Buckley, F.G. eds), Ornithological Monographs No. 36. American Ornithologists Union: Lawrence, Kansas, 85-112.

FHLC, (1998). Propuesta del plan de manejo ambiental para el humedal La Conejera y el bosque relictual Maleza de Suba. Bogotá, Colombia. Fundación Humedal la Conejera.

Guzmán, A. and Flórez, C. (2005). Monitoreo del proceso de restauración ecológica en los bordes e islas del Humedal Juan Amarillo (primera fase). Proyecto de investigación aplicada en la restauración ecológica del Humedal Juan Amarillo. Convenio de cooperación tecnológica Acueducto de Bogotá-Conservación Internacional.

Guzmán, A. (2012). Plantas de los humedales de Bogotá y el valle de Ubaté. Fundación Humedales. Instituto de Investigación de Recursos Biológicos Alexander von Humboldt. Fondo Hugo de Vries (Amsterdam). Bogotá, Colombia.

Hernández, J. and Rangel-Ch, O. J. (2009). La vegetación del humedal de Jaboque (Bogotá, D.C.). *Caldasia*, 31, 355-379.

Herrera, Y., Díaz, M.C., Vargas, P.L., Rodas, J.C. and Díaz, C.A. (2004). Política de Humedales del Distrito Capital de Bogotá: plan estratégico para su restauración, conservación y manejo. Bogotá.

Hilthy, S. L. and Brown W.L. (1986).A guide to the Birds of Colombia. New Jersey, U.S.A: Princeton University Press.

IDEADE-EAAB, (2009). Plan de Manejo Ambiental del Humedal de La Vaca. Pontificia Universidad Javeriana. Facultad de Estudios Ambientales y Rurales. Instituto de Estudios Ambientales para el Desarrollo IDEADE. Empresa de Acueducto y Alcantarillado de Bogotá.

Junk, W. J. (2012). Current state of knowledge regarding South America wetlands and their future under global climate change. *Aquatic Science - Research Across Boundaries*. DOI:10.1007/s00027-012-0253-8.

Miles, C. (1947). Los peces del Río Magdalena. Bogotá, Colombia: Editorial El Gráfico.

Muñoz, J.M. (2006). Estudio de Sedimentos Hídricos de Cinco Humedales de Bogotá y el sector de Campo Verde en la Localidad de Bosa- EAAB, Bogotá.

Molina, L. F. and Osorio, J. (1995). Guía de aves de Santa Fe de Bogotá. DAMA /AMSFB. Bogotá, Colombia: Gráficas de La Sabana Ltda.

Montenegro-S, A. L., Ávila Y. A., Mendivelso-Ch, H.A. and Vargas, O. (2006). Potencial del banco de semillas en la regeneración de la vegetación del humedal Jaboque, Bogotá, Colombia. *Caldasia,* 28, 285-306.

Moreno, V., García, J.F., Villalba, J.C., (2005). Descripción general de los humedales de Bogotá, D.C. Sociedad Geográfica de Colombia, 28 pp. http://www.sogeocol. com.co/documentos/humed.pdf.

Osbahr, K. and Gómez, N.C. (2011). abundancia, uso de hábitat y comportamiento de la Tingua Moteada (*Gallinula melanops bogotensis* Chapman 1914) En El Humedal Guaymaral, Bogotá – Colombia. *Revista U.D.C.A Actividades and Divulgación Científica*, 14, 81-91.

Ospina, D.M. (2007). Primera aproximación a la formulación de un plan estratégico para la conservación del Humedal Jaboque (Bogotá, D.C.). Universidad el Bosque. Bogotá. 107 p.

Palacio D.C. and Van der Hammen M.C. (2007). Redes heterogéneas del patrimonio. Los Casos del Centro histórico y el humedal Córdoba, Bogotá (Colombia). *Revista Hispana para el Análisis de Redes Sociales*, 13, 1-27.

Pinilla, G. (2010). An index of limnological conditions for urban wetlands of Bogotá city, Colombia. *Ecological Indicators*, 10, 848-856.

PUJ-EAAB (2007). Plan de manejo ambiental del humedal de Techo. Instituto de Estudios Ambientales para el Desarrollo IDEADE. Emresa de Acueducto y Alcantarillado de Bogotá.

Renjifo, L.M., Franco-Maya, A.M., Amaya-Espinel, J.D., Kattan, G.H., López-Lanús, B. (eds). (2002). Libro rojo de aves de Colombia. Serie Libros Rojos de Especies Amenazadas de Colombia. Instituto de Investigación de Recursos Naturales Alexander von Humboldt y Ministerio del Medio Ambiente: Bogotá.

Reynolds, C.S. (2006). The Ecology of Phytoplankton.London, U.K: Cambridge University Press.

Rivera-Rondón, C., Zapata, A., Valderrama, L., Paez, V., Baron, C., García R., Vélez, F. and Rondón, C. (2008). Evaluación de alternativa para el Control de Floraciones Algales en el Humedal Juan Amarillo. En: Experiencias de Restauración Ecológica en Colombia. (Barrera-Cataño J., Aguilar-Garavito D. and Rondón-Camacho, C eds.) pp. 233-248.

Rodríguez-Mahecha, J.V., Orozco-Rey, R.H., Vásquez-Vásquez, V.H., Calvachi-Zambrano, B., Mendoza-Sabogal, J., Alvarez-León, R., Camacho-Camacho, G., Hernández-Camacho J.I., and Smith-Mumm. U. (2000). Síntesis del estado actual de los humedales de la Sabana de Bogotá. Cap. 5 Santa María del Lago (pp. 64-72). Proyecto Recuperación de los Humedales de la Sabana de Bogotá: Alternativa hacia su Viabilidad Ecológica y Social. Bogotá, Colombia: Empresa de Acueducto y de Alcantarillado de Bogotá / Conservación Internacional-Colombia.

Rosselli, L. and Stiles, F. G. (2012). Wetland habitats of the Sabana de Bogotá Andean Highland Plateau and their birds. *Aquatic Conservation: Marine and Freshwater Ecosystems*, 22, 303-317.

Roth, R.A. (2009). Freshwater aquatic biomes. USA: Greenwood Press.

Sáenz-Guarín, J. E. (2001). Diagnóstico bioecológico del estado actual de la ictiofauna en el humedal Santa María del Lago, Bogotá: p. 59 In: Memorias Resúmenes VI Simposio Colombiano de Ictiología. ACICTIOS / UNC-DB / ACCEFN. Bogota D. C. (Colombia), Agosto 16-18, 60 p.

Sánchez-N, D. and Amát-García, G.D. (2005). Diversidad de la fauna de artrópodos terrestres en el humedal de Jaboque, Bogotá-Colombia. *Caldasia*, 27, 311-329.

Schmidt-Mumm, U. (1998). Vegetación acuática y palustre de la Sabana de Bogotá y Plano del Río Ubaté. Tesis de Maestría. Universidad Nacional de Colombia, Bogotá, Colombia.

SDA, 2008. Plan de Manejo Humedal de Córdoba. Alcaldía Mayor de Bogotá. Secretaría Distrital de Ambiente. Bogotá, 13p.

SDA, 2010a. Plan de Manejo del Humedal Juan Amarillo. Alcaldía Mayor de Bogotá. Secretaría Distrital de Ambiente. Bogotá. 17 p.

SDA, (2010b). Plan de Manejo Humedal La Vaca. Alcaldía Mayor de Bogotá. Secretaría Distrital de Ambiente. Bogotá. 15 p.

SDA, (2010c). Plan de Manejo Humedal El Burro. Alcaldía Mayor de Bogotá. Secretaría Distrital de Ambiente. Bogotá. 15 p.

SDA, (2010d). Plan de Manejo Humedal Capellanía. Alcaldía Mayor de Bogotá. Secretaría Distrital de Ambiente. Bogotá. 12 p.

Valencia, I. D. (1999). Registro de aves del humedal Santa María del Lago: 26 de Septiembre. Asociación Bogotana de Ornitología. Santa Fe de Bogotá D. C. Informe Técnico, 2 p.

Vasconcellos, P. C., Souza, D.Z., Ávila, S.G., Araújo, M.P., Naoto, E., Nascimento, K.H., Cavalcante, F.S., Dos Santos, M., Smichowski, P. and Behrentz,E. (2011). Comparative study of the atmospheric chemical composition of three South American cities. *Atmospheric Environment*, 45, 5770-5777.

Van Geel, B. and Van der Hammen, T. (1973). Upper Quaternary vegetational climate sequence of the Fuquene area (Eastern Cordillera, Colombia). *Palaeogeography, Palaeoclimatology, Palaeoecology,* 14, 9-92.

Van Der Hammen, T., Stiles F. G., Rosselli L., Chisacá M. L., Camargo, G., Guillot, G., Useche, Y. and Rivera, D. (2008). Protocolo de recuperación y rehabilitación ecológica de humedales en centros urbanos. Secretaría Distrital de Ambiente. Bogotá, D. C., Colombia.

Zapata-Anzola A., Rivera-Rondón C., Páez V., Pedraza, E. and García R. (2009). Factors Controlling Continual Cyanobacterial Blooms in a Tropical Urban Wetland. *Verhandlungen des Internationalen Verein Limnologie*, 30, 813-816.

In: Colombia
Editors: G. Lavigne and C. Cote
ISBN: 978-1-62417-848-1

Chapter 2

ELASMOBRANCHS FROM MARINE AND FRESHWATER ENVIRONMENTS IN COLOMBIA: A REVIEW

F. Ramírez[1]* ***and T. L. Davenport***[2]
[1]Pontificia Universidad Javeriana, Facultad de Ciencias, Bogotá, Colombia
[2]Director of Research and Development, Vivafresh Technologies, Florida City, United States of America

ABSTRACT

There is increasing interest in elasmobranch biology and diversity in the marine and continental environments of Colombia. The number of elasmobranch species are estimated to be about 124, but this number is likely to increase as more elasmobranchs are reported. Of these, eight species are freshwater stingrays. Populations of sharks and rays have been reducing in Colombia since the 1960s. Shark, ray and chimera fishery landings were about 157 T from 1958 to 1970 which reduced to about 76 T between 2001 and 2009 in the Colombian Atlantic Ocean. Shark, ray and chimera catches have diminished from 539 T from 1981 to 1990 to 353 T between 2001 and 2009 in the Colombian Pacific Ocean. However, catches from illegal fishing vessels of national and international origin are unreported so the actual rate of attrition is unknown. Broad taxonomic information has been gathered by the Colombian environmental and governmental institutions since 1957.

* E-mail: fermp44@yahoo.com., fernando.ramirez@javeriana.edu.co.

Little research has been conducted on elasmobranch feeding ecology in the Colombian Atlantic and Pacific Oceans when compared to the total amount of elasmonbranch species reported for Colombia. Most studies in marine environments have focused on demersal elasmobranchs. Elasmobranch reproduction has been reported off the Atlantic and Pacific coasts and on the Magdalena River. Shark movements have been studied off the Colombian Pacific Ocean. Evidence exists from both scientists and fishermen that sharks venture into Colombian rivers. Colombia recently developed a National Action Plan for Conservation and Management of sharks, rays and chimeras. Elasmobranch fisheries in Colombia are not well documented. According to FAO (Food and Agriculture Organization) fisheries statistics, both Pacific and Atlantic oceans show a decrease in elasmobranch landings since the 1960s. More research is needed by Colombian scientists to better understand elasmobranch biology in the waters of northern South America.

Keywords: Fisheries, South America, Potamotrygonidae, chimera, sawfish, conservation

INTRODUCTION

Colombia occupies 401,042 mi ^{2} (1,038,699 km ^{2}) of the north western end of South America. Its western shore is bordered by the Pacific Ocean, and its northern shore borders the Atlantic. Inland waters are characterized by rivers, streams, floodplains, lakes and lagoons. These habitats are important for elasmobranch species. Elasmobranch studies are in their infancy due to the lack of basic information on their ecology, population dynamics, feeding, conservation, and reproduction; however, recently (2000-present), an increasing number of investigations have been carried out on Colombian elasmobranch biology. These studies have focused mainly on marine elasmobranchs (Navia *et al.*, 2006a,b,c,d 2009, 2012; Mejía-Falla *et al.*, 2006, 2007; Acero *et al.*, 2007; Mejía-Falla and Navia, 2009; Grijalba-Bendeck and Acevedo 2009; Payán *et al.*, 2010a) and freshwater sting rays (Lasso *et al.*, 2010). Other investigations have focused on species native to Colombian rivers such as the Magdalena, Cauca, Catatumbo, Amazon, among others and their tributaries (Mojica-C, 1999; Lasso *et al.*, 2004, 2009; Mojica *et al.*, 2005, 2006; Villa-Navarro *et al.*, 2006; Maldonado-Ocampo *et al.*, 2006, 2008; Mejía-Falla *et al.*, 2009). The number of marine and freshwater elasmobranchs species studied is low compared to the total number of elasmobranchs reported for Colombia (124); however, there is a positive trend in the number of recent shark and ray investigations in both Pacific and Caribbean Oceans. Little is known about freshwater elasmobranchs as evidenced by few publications on the topic. Ecology and feeding habits have been examined using stomach content analyses and food web interactions (Gómez *et al.*, 2003; Mejía-Falla *et al.*, 2006; López *et al.*, 2009; Navia *et al.*, 2006a,d, 2007, 2009, 2010a, 2011, 2012; Payán *et al.*, 2011b) in the Colombian Pacific (Gómez *et al.*, 2003; Mejía-Falla *et al.*, 2006; López *et al.*, 2009; Navia *et al.*, 2006d, 2007, 2009, 2010, 2011; Payán *et al.*, 2011) and Caribbean Oceans (Téllez *et al.*, 2006; Polo-Silva and Grijalba-Bendeck, 2008; Grijalba-Bendeck *et al.*, 2009; Moreno *et al.*, 2009) as well as on freshwater sting rays (Mejía-Falla *et al.*, 2009). Length weight relationships (Mejía-Falla and Navia 2011a) and reproductive biology studies have focused on reproductive stages (immature and mature) in both females and males (Acevedo *et al.*, 2007; Grijalba-Bendeck *et al.*, 2008, 2009; Moreno *et al.*, 2008; Navia *et al.*, 2006d, 2009; Mejía-Falla *et al.* 2011), ray reproductive system description (Acero *et al.*, 2008) and embryo malformation (Mejía-Falla *et al.*, 2011). These studies were conducted in the Caribbean (Acevedo *et al.*, 2007; Grijalba-Bendeck *et al.*, 2008, 2009; Moreno *et al.*, 2008; Acero *et al.*, 2008) and Pacific Colombian Oceans (Navia *et al.*, 2006d, 2009; Mejía-Falla *et al.*, 2011; Payán *et al.*, 2010a). Freshwater sting ray reproductive information has focused on relatively few species (Mejía-Falla *et al.*, 2009; Argumentado, 2008). Migration studies have focused on the endangered, *Sphyrna lewini* (Griffith and Smith 1834) (Bessudo *et al.*, 2011). Fisheries information has been reported by several governmental, private, non-governmental and other institutions or organizations (García *et al.*, 2007; Grijalba-Bendeck *et al.*, 2008; Moreno *et al.*, 2008; Caldas *et al.*, 2009; FAO, 2011; Navia *et al.*, 2009, Navia and Mejía-Falla 2006a, 2010b). Broad taxonomic information related to fisheries has been gathered by Colombian environmental and governmental institutions since 1957 (Caldas *et al.*, 2009; Mejía-Falla and Navia 2011b). This information, however, is unreliable as elasmobranch catches are not discriminated at the species level or fishing zone (Caldas *et al.*, 2009). Other sources of fishery information have been provided by non-governmental organizations (Funadación MALPELO and SQUALUS) and studies by Mejía-Falla *et al.* (2006), García *et al.* (2007), and Grijalba-Bendeck *et al.* (2008). Funadación MALPELO and SQUALUS have provided fishing, economic and social related information for the Colombian Pacific Ocean (Mejía-Falla and Navia, 2011b). Most of these studies have focused on elasmobranch fisheries at specific ports along the Pacific but only recently since 2000. These reports provide data about elasmobranch species landings, economic and social related aspect to fisheries (Mejía-Falla and Navia, 2011b). Other fishery information has been collected mainly by FAO (FAO, 2011). A general decrease in the number of captures has been reported in both the Pacific and Atlantic waters off Colombia since the 1960s to the present (FAO, 2011). Elasmobranchs are often captured below their reproductive minimum size in the Pacific Ocean (Mejía-Falla and Navia, 2011b). This likely will eventually lead to population declines but further research is required. Less is known about the fishery status of elasmobranchs in the Colombian Caribbean Ocean. Apart from governmental statistics that are inconsistent and unreliable for the most part, there has been little published on the subject. Freshwater elasmobranch reports are unreliable estimates of fish populations since capture numbers are mostly provided by the ornamental fish industry. Based on the current elasmobranch fishery information, the government should take future actions to develop a consistent elasmobranch statistical and monitoring system. Conservation of Colombian elasmobranchs should have a high priority among governmental and environmental decisions. Recently, Colombia developed a conservation and management plan for all elasmobranch species (Caldas *et al.*, 2010). This plan is a key factor in improving protective measures as well as establishing priorities for elasmobranchs. About 50% of the total Colombian elasmobranchs have a data deficient category according to the IUCN (International Union for Conservation of Nature), and 25% are classified as nearly endangered. These estimates raise the question if there will be some elasmobranchs left for future generations. It is hoped that the recent conservation plan will result in action and, thus, provide the necessary regulations for thoughtful and reliable conservation of these marvelous creatures. It is hoped this chapter will provide a more detailed picture of the current level of knowledge of this important cartilaginous group.

MATERIALS AND METHODS

The current review is based on a literature search that included published information on elasmobranch diversity, feeding ecology, reproduction, population estimates, movements, conservation, fisheries and freshwater elasmobranchs in Colombia. Information is derived from articles published from 1940-2012 as few reliable articles were found prior to 1940. Published information was searched on databases such as: Biological Abstracts, ScienceDirect, SpringerLink, Scopus, EbscoHost, Annual Reviews, Scielo, Redalyc, ProQuest, Taylor and Francis, JSTOR and Whiley Online Library.

Recent published books were also included. Information from Red Colombiana de Investigadores de Condtrictios (Colombian Chondrychtyans Researcher Web) is available at http://www.squalus.org/redcondrictios /index.html and Grupo de Investigación en Peces de Caribe (GIPECA) (Caribbean Fish Research Group), and research results from Universidad Jorge Tadeo Lozano are available at: http://www.utadeo.edu.co /programas /pregrados/biolog_marina/gipeca/productos.php. FAO shark ray and chimera statistics data, analyzed from 1957 to 2009, is available at: http://www.fao.org /fishery/statistics/global-capture-production/en.

Elasmobranch species reports from Colombia´s waters were searched in FishBase (Froese, and Pauly, 2011). Other sources of elasmobranch species reports included selected publications (Navia *et al.*, 2006c; Mejía-Falla *et al.*, 2007; Acero *et al.*, 2007; Mejía-Falla and Navia, 2009; Grijalba-Bendeck and Acevedo 2009, Lasso *et al.*, 2010; Payán *et al.*, 2010b).

Diversity and Populations

Appearing during the Triassic period, Neoselachians make the totality of modern sharks and rays. Nearly 950 species of neoselachians exist today including 403 described shark-like species and 534 skates and rays (Nelson, 2006; Helfman *et al.*, 2009). Others report 1,100 existing neoselachian elasmobranchs (sharks and rays) and more than 30 species of holocephalans (chimaeras) (Compagno, 1990; Bone and Moore, 2008). Furthermore, in the last decade, chondrichthyan taxonomy has undergone a scientific renaissance with >180 new species formally described (White and Last, 2012). This effort encompasses *c.* 15% of the global chondrichthyan fauna, which consists of 1,185 currently recognized species (White and Last, 2012). Few reports or research projects have been conducted to estimate elasmobranch diversity in the marine or continental environments of Colombia. Froese and Pauly (2011) reported 92 species in Colombia grouped in 31 families (Table I). Others report 124 elasmobranch species determined from confirmed reports in Colombia (Mejía-Falla *et al.*, 2007; Acero *et al.*, 2007; Mejía-Falla and Navia, 2009; Grijalba-Bendeck and Acevedo 2009, Lasso *et al.*, 2010) (Table I).

Grijalba-Bendeck *et al.* (2009) report 49 shark and 25 ray species in the Colombian Caribbean Ocean. Navia *et al.* (2009) reported 35 confirmed shark and 24 ray species in the Pacific Ocean off Colombia.

Hleap *et al.* (2010) reported 19 shark and 13 ray species that occur at the Gorgona's National Park in the Colombian Pacific Ocean. Further exploration of marine environments will increase the current number of elasmobranch species reported for both the Pacific and Atlantic Oceans in tropical Colombia.

Only 5% of elasmobranch species live in freshwater compared to more than 40% of known teleost species (Ballantyne and Robinson, 2010). World estimates of at least 171 elasmobranch species, representing 68 genera and 34 families have been recorded from freshwater and estuarine environments (Martin, 2005).

Among these, 33 species are found in brackish waters, 91 in marginal, 13 in euryhaline, and only 36 are obligate to live in freshwater environments (Martin, 2005). Two thirds of the total worldwide freshwater elasmobranchs are Potamotrygonid river sting rays (Martin, 2005). Potamotrygonids can be distinguished from marine and freshwater stingrays of other families by the well-developed anteromedian projection on their pelvic girdle, called the prepelvic process (Garman 1877, 1913; Bigelow and Schroeder, 1953; Rosa *et al.*, 2010).

Potamotrygonidae is a family comprised of 22 valid species, 4 genera including *Potamotrygon*, *Paratrygon*, *Plesiotrygon* and the recently described genus *Heliotrygon* (Charvet-Almeida *et al.*, 2002; Rosa *et al.*, 2010; de Carvalho and Lovejoy, 2011). More species are likely to be further described as freshwaters in South America are explored. The total number of Potamotygonid stingray species reported in Colombia is ten (Table II), but this number is reduced to eight because species such as *Potamotrygon castexi* (Castello and Yagolkowski 1969) and *Potamotrygon hystrix* (Müller and Henle 1841) need further revision to confirm their identification (Mejía-Falla *et al.*, 2009). The Potamotrygonidae family has a problematic taxonomy and their identification is difficult e.g. *Potamotrygon motoro* (Müller and Henle 1841) might need revision among the Amazon morphs (Rosa *et al.*, 2010).

A thoughtful revision of the Potimotrigonid species of Colombia could possibly lead to more species described, but further research is required. World elasmobranch populations are declining (Clarke *et al.*, 2006; Robbins *et al.*, 2006; Myers *et al.*, 2007; Heithaus *et al.*, 2007; Ferretti *et al.*, 2008; Heithaus *et al.*, 2010). Chondrichthyans (cartilaginous fishes) include two subclasses, the elasmobranchs (shark-like fishes) and the holocephalans (chimaeras) (Helfman *et al.*, 2009). The greatest threat to chondrichthyan populations and species is attrition through fishing mortality (Dulvy and Forrest, 2010).

Over the past decade, there have been an increasing number of studies documenting declines of coastal and oceanic sharks and rays (Dulvy and Forrest, 2010).

Table I. Elasmobranch species from Colombia after FishBase (FB) and Mejía-Falla *et al.* (2007); Acero *et al.* (2007); Mejía-Falla and Navia (2009); Lasso *et al.* (2010) (M-F). (SN) Species Number, (FW) Freshwater

Family	SN FB	Percent (%)	SN M-F	Percent (%)	Distribution							
					Atlantic FB	Atlantic M-F	Pacific FB	Pacific M-F	Atlantic and Pacific FB	Atlantic and Pacific M-F	FW FB	FW M-F
Anacanthobatidae	1	1,04	2	1,60	1	2						
Alopiidae	3	3,13	2	1,60			1		2	2		
Carcharhinidae	19	19,79	19	15,20	4	5	5	4	10	10		
Centrophoridae	0	0,00	1	0,80		1						
Dasyatidae	6	6,25	6	4,80	4	3	2	3				
Echinorhinidae	1	1,04	0	0,00			1					
Etmopteridae	3	3,13	7	5,60	2	6	1	1				
Ginglymostomatidae	1	1,04	1	0,80					1	1		
Gymnuridae	2	2,08	2	1,60	1		1	2				
Heterodontidae	1	1,04	3	2,40			1	3				
Hexanchidae	0	0,00	4	3,20		2		2				
Lamnidae	2	2,08	1	0,80					2	1		
Mobulidae	0	0,00	2	1,60				2				
Mitsukurinidae	0	0,00	1	0,80	1							
Myliobatidae	8	8,33	1	0,80	3		1		4	1		
Narcinidae	4	4,17	7	5,60	2	4	2	3				
Odontaspididae	2	2,08	1	0,80	1		1	1				
Potamotrygonidae	3	3,13	10	8,00							3	10
Pristidae	4	4,17	1	0,80				1	4			
Pseudocarchariidae	1	1,04	0	0,00					1			
Rhincodontidae	1	1,04	1	0,80					1	1		
Rhinobatidae	6	6,25	5	4,00	1	1	5	4				
Rhinochimaeridae	1	1,04	0	0,00	1							
Rhinopteridae	0	0,00	3	2,40		2		1				
Scyliorhinidae	4	4,17	8	6,40	4	8						
Sphyrnidae	5	5,21	6	4,80			1		4	6		
Squalidae	0	0,00	1	0,80		1						
Squatinidae	1	1,04	2	1,60	1	1		1				
Torpedinidae	1	1,04	3	2,40		2	1	1				
Triakidae	4	4,17	7	5,60	3	4	1	3				
Rajidae	8	8,33	10	8,00	5	8	2	2	1			
Urotrygonidae	4	4,17	8	6,40	2	2	2	6				
Total	**96**	**100**	**125**	**100**	**36**	**52**	**28**	**40**	**30**	**22**	**3**	**10**

Table II. Freshwater stingrays of Colombia. After Mejía *et al.*, 2007; Maldonado-Ocampo *et al.*, 2008; Froese and Pauly, 2011; Lasso *et al.*, 2004, 2011. See text for detailed information about distribution. Unconfirmed species reports: *P. castexi and P. hystrix*

Species	Ditribution	IUCN Status
Paratrygon		
Paratrygon aiereba (Müller & Henle, 1841)	Amazon and Orinoco River Basins	Data deficient
Potamotrygon		
Potamotrygon castexi Castello & Yagolkowski, 1969	Amazon River Basin	Data deficient
Potamotrygon constellata (Vaillant, 1880)	Amazon River Basin	Data deficient
Potamotrygon hystrix (Müller & Henle, 1841)	Amazon River Basin	Data deficient
Potamotrygon magdalenae (Duméril, 1865)	Endemic: Magdalena, Atrato and Cauca River Basins	Near Threatened
Potamotrygon motoro (Müller & Henle, 1841)	Amazon, and Orinoco River Basins	Data deficient
Potamotrygon orbignyi (Castelnau, 1855)	Orinoco River Basin	Least concern
Potamotrygon schroederi Fernández-Yépez, 1958	Amazon, and Orinoco River Basins	Data deficient
Potamotrygon yepezi Castex & Castello, 1970	Colombian Atlantic Rivers	Data deficient
Plesiotrygon		
Plesiotrygon iwamae Rosa, Castello & Thorson, 1987	Amazon River Basin	Data deficient

Elasmobranch population estimates from Colombia are few. Most information has been recorded by FAO as annual fishery catches (FAO, 2011). Reduced shark and ray biomass harvest was reported between 1970 to 2001 in the Colombian offshore Caribbean Sea (García *et al.*, 2007). Commercial fishing negatively impacted shark and ray populations during 2005 to 2008 in the Caribbean Sea near San Andrés, Providencia and the Santa Catalina Archipiélago in Colombia (Castro-González and Ballesteros-Galvis, 2009).

Large pelagic sharks, including hammerheads (*Sphyrna* spp.), blue shark *Prionace glauca* (Linnaeus 1758), porbeagle shark *Lamna nasus* (Bonnaterre 1788), shortfin mako *Isurus oxyrinchus* (Rafinesque 1810), and thresher shark *Alopias vulpinus* (Bonnaterre 1788) are nearly extinct (Ferretti *et al.*, 2008).

Most individuals captured were juveniles (Castro-González and Ballesteros-Galvis, 2009). Moreover, shark and chimera fishery landings were about 157 T between 1958 and 1970 but had a reduction to about 76 T between 2001and 2009 in the Colombian Atlantic Ocean (Figure 1) (FAO, 2011). Reduced elasmobranch (shark and ray) size ranges and landings have been reported to occur from 1994 to 2004 off the Colombian Pacific coast (Zapata *et al.,* 1998; Caldas *et al.*, 2009). Moreover, shark and ray biomass has been reduced from 539 T between 1981-1990 to 353 T between 2001-2009 in the Colombian Pacific Ocean (Figure 2) (FAO, 2011); however, unreported catches from illegal fishing vessels of national and international origin are unknown.

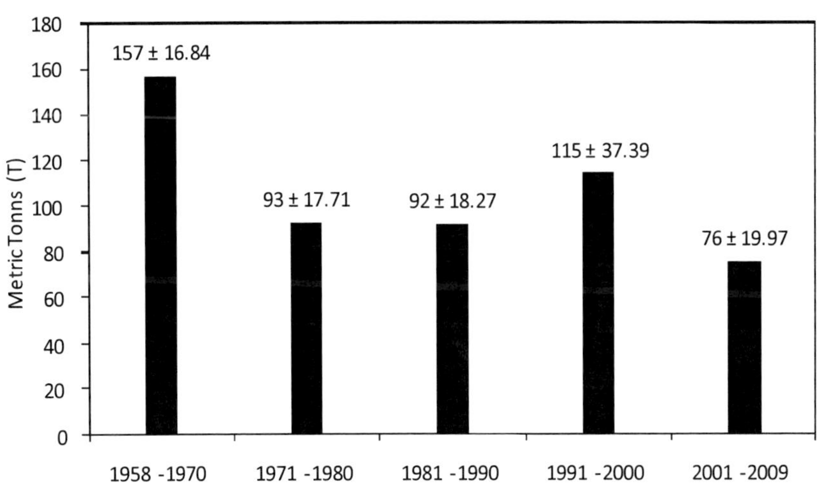

Figure 1.

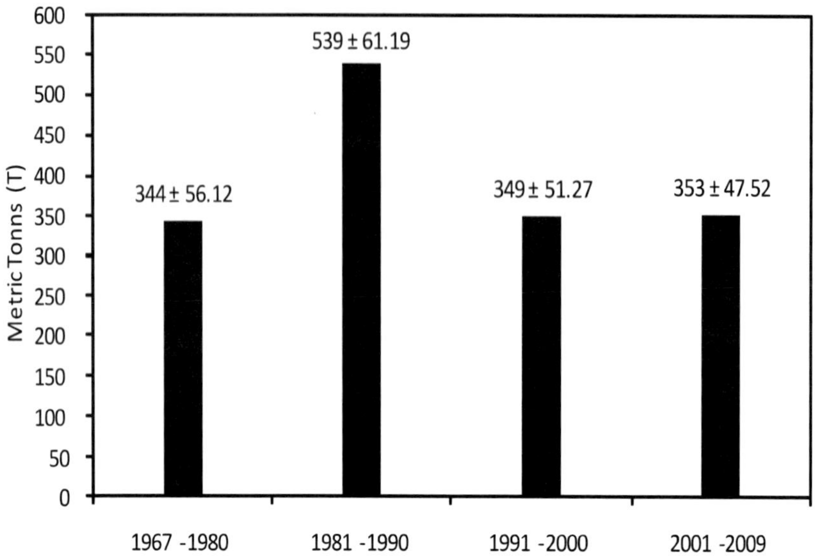

Figure 2.

Reliable population estimates are influenced by unreported catches. Majía-Falla and Navia (2006) studied the population dynamics of *Triaenodon obesus* (Rüppell 1837), *Rhincodon typus* (Smith 1828) and *Manta birostris* (Walbaum 1792) at four zones of Gorgona's National Park in the Colombian Pacific Ocean. They reported that most sightings were recorded in the northern zone. Southern and western zones were occupied mainly by *T. obesus* with female-male groups including juveniles.

Freshwater stingray population estimates are few, and most information is provided by ornamental fish exploitation; however, the numbers of official stingray catches have been proposed to be low because rays have a high price per unit in contrast to teleost ornamental fishes. Official live stingray specimen numbers were reported to be 28.686 units (0.15% of the total ornamental fish capture) in 1996 (INPA, 1997).

Feeding Ecology

Recent research has focused on the feeding ecology of elasmobranchs off the Pacific coast of Colombia (Gómez *et al.*, 2003; Mejía-Falla *et al.*, 2006; Navia *et al.*, 2006d, 2007, 2010a, 2011, 2012; López-García et al., 2012). Most of these studies concentrated on benthic and demersal elasmobranchs.

Elasmobranch species, such as the sharks: *Carcharhinus leucas* (Müller and Henle 1839, *S. lewini*, *Carcharhinus limbatus* (Müller and Henle 1839), *Galeocerdo cuvier* (Péron and Lesueur 1822) and *Carcharhinus falciformis* (Müller and Henle 1839); the rays *Aetobatus narinari* (Euphrasen 1790) *Urotrygon rogersi* (Jordan and Starks 1895) and *Raja velezi* (Chirichigno 1973); the guitarfishes *Zapteryx xyster* (Jordan and Evermann 1896) and *Rhinobatos leucorhynchus* (Günther 1867) are predatory. They have high and intermediate trophic levels (3 to 4.5) in the Gulf of Tortugas off the Colombian Pacific shore between Guapi Inlet and Buenaventura Bay (2°45'N, 78°10'W to 3°50'N, 77°20'W) (Navia *et al.*, 2010a). These elasmobranchs reside in benthic and demersal habitats, thus feeding mainly on fishes and invertebrates of the benthic epifauna and endofauna (Navia *et al.*, 2010a). Moreover, in the Gulf of Tortugas, using computer tropic networks analyses; Navia *et al.* (2012) were able to identify complex food web interactions within various scenarios. Scenario 1, representing trophic networks left unaltered with 250 nodes and 579 interactions among elasmobranchs, teleosts, invertebrates, zooplankton and phytoplankton. Scenario 2, shrimps removed from the trophic networks which generated 232 nodes and 398 interactions. Scenario 3, shrimps and snappers removed generating 229 nodes and 349 interactions. Scenario 4, other organisms removed 226 nodes and 329 interactions (Navia *et al.* 2012). Among feeding studies, Navia *et al.* (2011), reported that *U. rogersi* (Elasmobranchii: Myliobatiformes) is a predator that feeds mainly on benthic fauna, including organisms that are attached to surface sediments (epifauna) and those within them (infauna), from the central zone of the Eastern Tropical Pacific off the Colombian coast and offshore islands, La Barra, El Tigre, Punta Soldado and Santa Barbara-Raposo (Navia *et al.*, 2011). *U. rogersi* has a heterogeneous diet based on a wide variety of epibenthic invertebrates, but with a high preference for shrimps (Navia *et al.*, 2011; Hleap *et al.*, 2012). It is a specialist species with a high preference for decapod shrimps. Decapods were the dominant food source in their diet providing 79.5% of the total consumed mass (% M), with 49.8% of the total percent number of items consumed (%N) and this item had a occurrence percent (%O) among other items in the stomach contents of 52.1%. Among the decapods, *Ogyrides alphaerostris* had the greatest contribution to the diet (13.6%N and 12·3%O), followed by Pasiphaeidae (5.9%N and 6.5%O). Polychaetes were also important contributors to their diet with (30.9%), mass (6.8%) and occurrence (40.8%). Among these, the families Nereidae (6.4%N and 8.9%O) and Onuphidae (4.8%N and 8.0%O) were important in the diet (Navia *et al.*, 2011). The relationship between size class and prey suggests that juveniles focus their foraging effort on infaunal organisms perhaps because of movement limitations, the palatability of these prey, or to reduce competition with larger, more skillful predators (Navia *et al.*, 2011). Moreover, individuals from size classes II (20 to <28 cm in total lenght) and III (28 to 36.5 cm in total lenght) preyed more upon decapods and stomatopods, suggesting that these growth stages focus on epifaunal prey (Navia *et al.*, 2011).

The diet of elasmobranchs, such as the houndshark *Mustelus lunulatus* (Jordan and Gilbert 1882), the rays, *Dasyatis longa* (Garman 1880), *R. leucorhynchus*, *R. velezi* and *Z. xyster*, were studied from the central fishing zone in the Pacific Ocean of Colombia (Navia *et al.*, 2007). All the studied species of elasmobranchs feed on epibenthic invertebrates (Navia *et al.*, 2007). These include benthic infauna (prey species living in the sediment) and epibenthic fauna (prey species living on the surface of sediments) (Navia *et al.*, 2007). *M. lunulatus* fed mostly on stomatopods (mantis shrimps) of the family Squillidae (73.2%N, 36.1W %, and 78.6%O). Within this family, *Squilla panamesis* was the main prey item with a 92.8 % index of relative importance, (I_{RI}). This index combines three separate measures (frequency, number and volume) of prey importance allowing a more representative feeding habit analysis (Caillet *et al.*, 1996). *S. panamesis* is followed by *Squilla parva* (9.27 % N, 5.95 % W, I_{RI} 4.01, 14.28 % O). Additional, less-represented prey categories included Decapoda, fishes, Gastropoda and Bivalvia (Mollusca). *M. lunulatus* feeds opportunistically on shrimps, crabs and gastropods in low percentages (Navia *et al.*, 2007). Navia *et al.* (2006d) studied the feeding habits of *M. lunulatus* in the central zone off the Colombian Pacific coast. The guts of individuals captured in the trawl net of a commercial shrimping vessel were examined (Navia *et al.*, 2006d). Three out of fifty stomachs were empty. The diet of *M. lunulatus* comprised 13 items. The most abundant items in the diet were the stomatopods (*S. panamensis* and *S. parva*). Other less frequently consumed species were: *Pseudosquilla similix*, *Euphilax* sp., *Portunus asper*, *Callinectes toxotes*, *Hepatus* sp., *Pennaeus occidentales*, *Trachypennaeus* sp., Palaemonidae, Hippidae, Terebridae, Buccinidae (Navia *et al.,* 2006d). Similar results were reported by Gómez *et al.* (2003) at the Gorgona's National Park, (2°58'10" N and 78°11'05" W) Colombian Pacific Ocean between June, 1988 and July, 1992 for *M. lunulatus* that fed mainly on shrimps (Crustaceae), such as *Portunus iridiscens* (41% by weight and 34% frequency), followed by *S. panamensis* (12% by weight), other invertebrates < 5% *Hypoconcha panamensis*, *Squilla mantoidea*, *Iliacantha hancoki*, *Munida*, *Calappa convexa* and fish remains (Gómez *et al.*, 2003). Another species of the same genus, *Mustelus henlei* (Gill 1863) studied off Gorgona Island, had a diet composition mainly of *S. panamensis* (34% by weight and 22% frequency) followed by fish remains (16.31% by weight and 6% frequency) and mollusks of the family Loliginidae (11% by weight and 8% frequency) (Gómez *et al.*, 2003). López *et al.* (2009) reported that Penaeidae (49.01%), were a common food item followed by crabs of the species, *Raninoides benedicti* (10.05%), in the gut of *D. longa* cought in the central fishing zone of the Pacific Ocean off the Colombian coast. Other prey items included decapods, gastropods and fishes (Navia *et al.*, 2007). Other studies carried near Buenaventura, throughout the central zone of the Pacific Coast of Colombia found that *D. longa* showed a narrow niche breadth feeding mainly on shrimps and fishes and its diet was dependent on size, but not on sex or season (López-García *et al.*, 2012). Juvenile individuals (class I) fed on shrimps, sub-adults (class II) on fishes and adults (class III) on stomatopods (López-García *et al.*, 2012). Other species such as *R. leucorhynchus* fed mostly on decapods (66.6 % O, 85.6 % N, 41.7 % W and 87.0% I_{RI}) followed by fish. The main prey species was *Trachypenaeus* sp. (60.4% I_{RI}). Stomatopods and crabs (Portunidae) were also found in its diet (Navia *et al.*, 2007). Recently, Payán *et al.* (2011) reported that *R. leucorhynchus* fed mostly on Penaeidae shrimps, especially *Trachypenaeus* sp. in the central fishing zone off the Pacific-Colombian coast (5°05'N, 77°16'W-2°31'N, 78°35'W). This study reported that females fed more on fishes and males fed on stomatopods and crustaceans. Navia *et al.* (2007) reported that *R. velezi* fed mostly on fishes (38.4% O, 38.5% N, 37.4% W, and 61.17% I_{RI}). *Z. xyster* fed most frequently on fishes and penaeid shrimps (> 16%). For this species, decapods were the most important prey item with 43.2 % I_{RI} (Navia *et al.*, 2007). Similarly, Mejía-Falla *et al.* (2006) reported that *Z. xyster's* diet was based mainly on the family Penaeidae (Crustaceae) (19.4% O, 27.3 % N 31.5% W and 11.26% I_{RI}) followed by Perciform fishes (16.1 % O, 27.0 % N 14.6 % W and 6.67% I_{RI}) and the crustacean *S. panamensis* (12.9 % O, 18.2% N, 26.1% W and 5.60% I_{RI}) in the central Pacific fishery zone south of Buenaventura Bay to Ensenada de Gapi (Mejía-Falla *et al.*, 2006).

The life history and ecology of most elasmobranchs from Colombia are poorly known. Recently, Grijalba-Bendeck *et al.* (2009) provided a concise review on shark and ray biological and ecological information, however, most of this information comes from theses, technical reports and unpublished results. Little research has been published in refereed journals on elasmobranch feeding ecology in the Colombian Atlantic Ocean. Recently, a number of studies have targeted the feeding biology of a few elasmobranch species (Téllez *et al.*, 2006; Polo-Silva and Grijalba-Bendeck, 2008; Moreno *et al.*, 2009) Polo-Silva and Grijalba-Bendeck (2008) studied the trophic spectrum of the guitar fish *Rhinobatos percellens* (Walbaum 1792) at Santa Marta, Caribbean Ocean. They found that *R. percellens* fed mainly on crustaceans of the Callianassidae (24.49 % O, 65% N, 0.53 W and 1801% I_{RI}) and Penaeidae (9.18 % O, 11 %N, 17.07 W 208 %I_{RI}) families. Other prey items were fishes that included: Haemulidae sp., Sciaenidae sp., and Clupeidae. *R. percellens* is considered a specialist predator during early developmental stages and a generalist when attaining sexual maturity. Moreno *et al.* (2010) studied the trophic spectrum of the electric ray *Narcine bancroftii* (Griffith and Smith 1834) (Elasmobranchii, Narcinidae) off Salguero Beach, Santa Marta, Colombian Caribbean. They reported that *N. bancroftii* fed mostly on Sipunculidae, Callianassidae and Ophichthidae. The females showed higher prey diversity values, and males had the highest uniformity values. The diet changed ontogenetically, and feeding areas were common to both sexes and all stages of development (Moreno *et al.*, 2010). Dahl (1971) provided information about the feeding ecology of sharks in the Colombian Atlantic Ocean. Dahl (1971) reported that the shark, *Sphyrna tiburo* (Linnaeus 1758), feeds on crabs, shrimps and other crustaceans. The Mako shark, *I. oxyrinchus* feeds on fishes, such as mackerel and snook. *Sphyrna zygaena* (Linnaeus 1758) feeds on crabs, shrimps, squids, other sharks, such as Ariidae, Scombridae and *Oligoplites* spp. (Carangidae). The blue shark, *P. galuca,* feeds on tunids (mackerel and bonitos) and squid. *Charcharias taurus* (Rafinesque 1810) feeds on fishes, octopus, squid and crabs (Dahl, 1971). Ray feeding ecology information has been provided by Dahl (1971).

The Venezuelan round stingray, Urotrygon venezuelae (Schultz 1949), feeds on crustaceans, worms and mollusks, Peneid shrimps, crabs and unidentified organic detritus (Dahl, 1971; Téllez et al., 2006). The cow nose ray, Rhinoptera bonasus (Mitchill 1815), feeds on crabs, crustaceans and filter-feeding invertebrates (Dahl, 1971).

Our knowledge of the life history and ecology of most freshwater and euryhaline elasmobranchs are limited (Martin, 2005). The diet of potamotrygonids varies among species and geographically within species (Rosa *et al.*, 2010). Otherwise, there are few studies about feeding ecology of freshwater stingrays found in Colombia. Recently, Mejía-Falla *et al.* (2009) reviewed the ecology of Colombian freshwater sting rays, finding no information about the feeding ecology of *Paratrygon aiereba* (Müller and Henle 1841), *Potamotrygon constellata* (Vaillant 1880), *P. motoro*, *Potamotrygon obrignyi* (Castelnau 1855), *Potamotrygon shroederi* (Fernández-Yépez 1958) or *Plesiotrygon iwamae* (Rosa Castello and Thorson

1987) in Colombia (Mejía-Falla et al., 2009). Feeding ecology studies can be found in theses, unpublished reports and internal reports (Mejía-Falla et al., 2009).

Most feeding ecology studies on the cited rays have been reported in Brazil and Venezuela (Mejía-Falla *et al.*, 2009). Others, (Salinas and Agudelo, 2000; Dahl, 1971; Galvis *et al.*, 1997 and Villa-Navarro, 1999) have reported freshwater ray feeding ecology information from Colombia. *P. hystrix* feeds on fishes, crustaceans and insects in the Amazon, Putumayo, Caquetá and Guaviare rivers (Salinas and Agudelo, 2000). *Potamotrygon magdalenae* (Duméril 1865) feeds manly on mollusks, other invertebrates and small fishes (Dahl, 1971). Galvis *et al.* (1997) reported that this species feeds on insect larvae and detritus at Catatumbo, Colombia. Villa-Navarro (1999) provided evidence that *P. magdalenae* feeds on invertebrates and is considered a detritivore at Prado Dam, Tolima State, Colombia. *Potamotrygon yepezi* (Castex and Castello 1970) feeds on insect larvae and detritus (Galvis *et al.*, 1997). Most of these feeding observations use broad taxonomic categories to identify prey items that over-generalize and do not provide detailed evidence on particular pray items at lower taxonomic levels.

REPRODUCTION

Elasmobranch reproduction is either oviparous, viviparous ovoviviparous or placental viviparous. Oviparous species are characterized by egg-laying, as is the case in some sharks and skates (Helfman *et al.*, 2009). Viviparous species, i.e. fishes, are born live as juveniles, having bypassed a free larval stage (Helfman *et al.*, 2009). This reproductive mode occurs in about half of the 1,200 species of Chondrichthyes. Ovoviviparity is characterized by young development inside the mother but with a dependency on the yolk that was laid down during oogenesis (Helfman *et al.*, 2009). Placental viviparity occurs in the families, Charcharhinidae and Sphyrnidae. Yolk is consumed and then the spent yolk sac attaches to the uterine wall to form a yolk-sac placenta (Helfman et al., 2009). Many elasmobranchs have internal fertilization but lay eggs that develop over many months outside the female (Helfman *et al.*, 2009). Sharks with internal fertilization produce relatively few, large young over a long developmental period. Newborns are small replicas of the adults and experience no larval stage. Gestation periods in sharks average 9 to 12 months. Rays have a yolk-sac viviparity as their mode of reproduction

(Musick and Ellis, 2005). In the stingrays (Myliobatoidei), the uterus develops large villous projections termed trophonemata that are important in respiratory and nutrient exchange with the developing embryo (Musick and Ellis, 2005). Some rays (for example, stingrays) have only one functional oviduct, and one copulation by one male may suffice to fertilize all of the eggs (Pratt and Carrier, 2005). Others have both functional oviducts (Teshima and Takeshita, 1992). Pregnancy only lasts several months in most rays compared to viviparous sharks; however, exceptions exist (Lutton *et al.*, 2005).

Elasmobranch reproduction has been examined off the Atlantic coast of Colombia. Recently (2000-2011), there has been an increasing trend in the number of studies on their reproductive biology (Grijalba-Bendeck *et al.*, 2008, 2009; Moreno *et al.*, 2008; Acero *et al.*, 2008). Grijalba-Bendeck *et al.* (2009) reviewed studies from theses, unpublished reports, refereed journals and internal reports. They concluded that little information has been published in international journals and that little is known about reproductive strategies among elasmobranchs off the Colombian Atlantic Ocean. Few reliable publications report on shark reproductive biology. Colombia is home to two of the world's smallest shark species. *Etmopterus perryi* (Springer and Burgess 1985) attains sexual maturity at 160 to 175 mm in males and at 190 to 200 mm standard length (*Ls*) in females (Springer and Burgess, 1985). The second species, *Etmopterus carteri* (Springer and Burgess 1985), reaches sexual maturity at 180 to 212 mm *Ls* in males. Both shark species are endemic to a narrow stripe off the Caribbean coast of Colombia between Barranquilla and Santa Marta (Springer and Burgess, 1985).

Moreno *et al.* (2010) examined the reproductive biology of the electric ray *N. bancroftii*, caught as bycatch in seines on two beaches of El Rodadero, Santa Marta on the Colombian Caribbean between August 2005 and October 2006. All five reproductive stages occurred in females; however, mature females without embryos (stage IIIa) were most commonly caught between August 2005 and February 2006. Fewer proportions of mature females were observed from March to May 2006. Females with embryos (IIIb) were observed from February to April. Females that already gave birth to embryos (stage IV) occurred from March to May of 2006. Reproductive stage II females occurred in fewer numbers during August and September of 2005 and in July, August and October of 2006. All reproductive stages were observed on males. The most frequently observed stage was mature (III), followed by immature (I) and those in the process of maturation (II). The gonado-somatic index (I_G), an index for comparing gonadal states (as between seasons and size classes) (Caillet *et al.*, 1996), increased during September to November. The maximum reproductive time in females was linked to the occurrence of more males during December (Moreno *et al.*, 2010). Females carried up to 14 embryos. *N. bancroftii* had one reproductive maximum per year in November followed by copulation. Gestation period was estimated around four months. Sexual maturity was attained at 32.2 cm *Ls* in females and between 20.2 to 22.9 cm *Ls* in males (Moreno *et al.*, 2010). Acero *et al.* (2008) studied the reproductive systems of four rays in the Caribean Ocean. *R. percellens* has a generalized female reproductive system, *U. venezuelae* and *Dasyatis guttata* (Bloch and Schneider 1801) have ovaries that include unique internal structures. *N. bancroftii* has a highly derived reproductive system, with uniquely shaped internal structures. Male reproductive systems were similar; however, testicles in *R. percellens* had some degree of external lobulation. Radial testicles occurred in *U. venezuelae* and *D. guttata* (Acero *et al.,* 2008). Acevedo *et al.* (2007) studied the reproductive aspects of *N. bancroftii, Diplobatis guamachensis* (Martin 1957) and *R. percellens*. Female *N. bancroftii* showed immaturity and spawned states, while the males were mature during the second semester of 2004. Female *D. guamachensis* had embryos, and males had mature testes. Female *R. percellens* were mature with embryos and spawned states, and males were mature (Acevedo et al., 2007).

The guitarfish *R. percellens* (Batoidea: Rajiformes) has a reproductive system that is similar to that of other rays studied at Salguero Beach near the Guaira River at Santa Marta off the Colombian Atlantic coast (Grijalba-Bendeck *et al.*, 2008). *R. percellens* produces a relatively low number of embryos (two to four) during each reproductive cycle (Grijalba-Bendeck *et al.*, 2008). It is a viviparous matrotrophic ray, with compound ovaries. Vitellogenetic development was simultaneous with embryonic growth. The first stages of the reproductive cycle start within mature females without mature ova, uterine-eggs or embryos from May to September. The reproductive period that includes mating, gestation and births occurred during October and April, and a reduction in activity was noted from May to September when embryos stopped their growth (Grijalba-Bendeck *et al.*, 2008). Their study suggests that copulation takes place during both December and February-March.

Recently, elasmobranch reproductive biology has attracted increasing research interest off the Pacific coast of Colombia (Navia *et al.*, 2006d, 2012; Payán *et al.*, 2010a, 2011; López García *et al.*, 2012; Mejía-Falla *et al.*, 2012). Reproductive stage I (immature, i.e. no white-cream colored ovocites present and oviducts had a filament-like form), was the most common stage (52.4%) of all captured females ranging in size from 54 to 66 cm *Ls* of *M. lunulatus* at the central fishing zone in the Pacific Ocean off Colombia. Stage II, (claspers do not reach the posterior end of pelvic fins and very thin gonads) was the most frequent stage (47.6%) of all captured males ranging in size between 58.4 and 75.0 cm *Ls* (Navia *et al.*, 2006d). Mean fecundity were six offspring per female ranging in size from 14 to 23 cm *Ls* (Navia *et al.*, 2006d). Sexual dimorphism was not observed to occur (Navia *et al.*, 2006d). Payán *et al.* (2011) reported that in *R. leucorhynchus*, half of the captured males (51.5%) developed calcified claspers, and half of all females (56.5%) were pregnant with 1 to 6 embryos between 3 to 19.5 cm. These observations suggested a birth size of 19 to 19.5 cm full-length (*Lf*) in the Colombian Pacific Ocean. Payán *et al.* (2010a) studied the reproductive biology of *Narcine leoparda* (Carvalho 2001) near Juanchaco in the Colombian Pacific Ocean. They found that males and females had reproductive stages I and II. Female *N. leoparda* ovocites (stage II) ranges between 0.49 cm y 2.29 cm and one considered between 1.7 and 4.5 cm.

The length of mature male calcified claspers were between 2.0 and 4.0 cm. Mejía-Falla *et al.* (2011) reported morphological abnormalities in embryos of *U. rogersi* in the Colombian Eastern Pacific. Two embryos had incomplete, deformed pectoral fins separated from the head along the anterior margin (Mejía-Falla *et al.* 2011). Furthermore, Female *U. rogersi* reached greater maximum total length (*L*T), disc width (*W*D) and mass (*M*) (38.0 cm, 19.9 cm and 348 g) than males (32.5 cm, 17.0 cm and 165 g) in the central-south Pacific coast of Colombia (3◦ 56'N; 77◦ 25' W to 3◦ 00' N; 77◦ 10'W) (Mejía-Falla *et al.*, 2012). Male first maturity was reached at 61.8% of máximum *W*D, and estimated *W*D50 was between 11.5 and 11.8 cm (Mejía-Falla *et al.*, 2012). Female size at first maturity was 52.8% of maximum *W*D, and estimated *W*D50 was between 11.8 and 12.3 cm (Mejía-Falla *et al.*, 2012). *U. rogersi* has a reproductive strategy based on low fecundity, a rapid reproductive cycle (short ovulation and gestation time), three birth peaks per year and large embryos (Mejía-Falla *et al.*, 2012). Mejía-Falla *et al.* (2006) reported that stage II was the most common gonadal reproductive stage among *Z. xyster* females (40.7%) in the central fishing zone off the Colombian Pacific Ocean.

All species of potamotrygonid stingrays studied to date adhere to a matritrophic aplacental viviparous reproductive mode with uterine villi termed trophonemata, which nourish the embryo during gestation (Araújo *et al.*, 2004). The gestation period of river stingrays varies enormously among species, ranging from three to 12 months; uterine fecundity ranges from one to eight pups per litter with individual females giving birth in alternate years (Araújo *et al.*, 2004; Charvet-Almeida *et al.*, 2005; Martin, 2005; Rosa *et al.*,

2010). The life histories of potamotrygonid stingrays are biologically constrained to a strongly K-selected pattern (Compagno and Cook, 1995) i.e. characterized by a high age at maturity, a low number of offspring and fecundity, and a lower level of mortality (Adams, 1980). Mejía-Falla *et al.* (2009) recently reviewed the reproductive biology of freshwater stingrays among other topics in Colombia. This review was based on unpublished reports, theses, internal reposts, books and refereed journal information. Little information has been reported on freshwater stingray reproductive biology. Mejía-Falla *et al.* (2009) reported that no reproductive biology information has been documented for *P. aiereba*, *P. constellata*, *Potamotrygon orbignyi* (Castelnau 1855), *Potamotrygon schroederi* (Fernández-Yépez 1958), *P. yepezi* in Colombia. Argumentado (2008) reported that female *P. motoro* reach sexual maturity between 27 to 31 cm of diskwidth. Rays have a broad surface on their dorsal and ventral portions of the body referred as the disk. The disk width is the length of widest point on the dorsal side of the ray′s disk Teshima and Takeshita (1992) studied the reproductive biology of the freshwater stingray *P. magdalenae* from lakes near San Cristobal, a small village located about 150 southeast of Cartagena, Colombia during August 12-19, 1980. Their study reported 117 rays captured; 44 females and 73 males. Oogenesis in *P. magdalenae* differs from that of other elasmobranchs in that follicular epithelial folds extend inward into the oocyte and enlarged cells were formed in the extending follicular epithelium (Teshima and Takeshita 1992). The corpus-luteum-like tissue is thought to originate from the follicular epithelial folds that are left in the postovulatory follicle. The large oocites found in the ovary were 25 mm in diameter. They double in diameter (50mm) when they enter the common ostium. Females appear to reach or have reached sexual maturity between 170 and 210 disk widths as evidenced by oocites of about 25 mm in diameter and embryos. During spermatogenesis in males of *P. magdalenae*, seminiferous follicles move from the central area toward the periphery of testicular masses as spermatogenic cells develop (Teshima and Takeshita, 1992). Males become sexually mature between 170 and 190 mm disk widths. Their study suggests that reproductive activity including ovulation, mating, fertilization, and parturition might take place over an extended period of time. Size at birth was estimated to be about 100 mm disk width (Teshima and Takeshita, 1992); however, caution must be taken when interpreting their results on reproductive activity because they only sampled individuals over an eight-day period and failed to capture individuals on a broader time scale. Others found that maturity is reached by *P. magdalenae* at a size of 240mm disk width by females and at 202 mm by males, while the minimum maturity size was 164mm (females) and 160mm (males) in Ciénaga de Solayo, Magdalena State, Colombia (Ramos-Socha and Grijalba-Bendeck *et al.*, 2011). The right ovary of females was found to be less developed than the left one, both uteri were functional containing 1 to 5 embryos. Birth size was estimated between 87 and 95mm (Ramos-Socha and Grijalba-Bendeck *et al.*, 2011). More research is required to determine whether or not *P. magdalenae* has a seasonal reproductive cycle at a particular time of the year as is typical of most. For example, reproductive cycles of freshwater stingrays, *P. motoro*, *P. orbignyi*, *P. schroederi*, *Potamotrygon scobina* (Garman 1813), *Potamotrygon* sp., *P. aiereba* and *P. iwamae* are directly related to the hydrologic cycle and have defined reproductive periods (Charvet-Almeida *et al.*, 2005).

MIGRATIONS AND FISH MOVEMENTS

Sharks are highly mobile in marine environments. Migrations are well documented (Munoz-Chapuli, 1984; Casey and Kohler 1992; Kohler *et al.*, 1998; Wilson *et al.*, 2006; Aires-da-Silva *et al.*, 2008; Domeier and Nasby-Lucas 2008; Goldman and Musick 2008; Kohler and Turner 2008; Stevens, 2010). Shark movements have been studied in the Pacific Ocean off Colombia. Bessudo *et al.* (2011) reported that endangered (IUCN) hammerhead sharks *S. lewini* ultrasonic tagged off the coast of Malpelo Island (3°58′N and 81°37′W) in the Colombian Pacific were detected near Cocos and the Galapagos Islands. Only five sharks out of 69 tagged at Malpelo were recorded at Cocos Island (627 km from Malpelo) from 2006 to 2009 (Bessudo *et al.*, 2011). None of the sharks were reported only at Galapagos Islands. One shark was detected both at Cocos and the Galapagos (710 Km from Cocos island) islands. Thus, the total distance covered by these sharks was *c.a.* 1,941 km from Malpelo through Cocos and the Galapagos. These observations support the hypothesis that hammerheads that occur at Malpelo are not exclusive to the island. Furthermore, their data provide evidence that hammerheads are leaving the island in March and April to move to birthing areas away from Malpelo and returning to the island at a later time. Most hammerheads visit Malpelo waters for extended periods of time throughout the year, then leave and return to the island after a couple of days or several months. Hammerheads swam at shallow depths of about 15 m depending on the time of the year and were found below, but also within and above the thermocline. They were, thus, not restricted to thermal gradients (Bessudo *et al.*, 2011).

Sharks and rays are primarily marine organisms, but a number of species regularly occur in fresh water including several taxa that are known only from fresh and brackish waters (Nelson, 2006). The spatial behavior of most sharks and rays that live or venture into freshwaters is poorly known, and well defined migrations do not seem to occur (Lucas *et al.*, 2001). Most aspects of saltwater sharks venturing into Colombian rivers are largely unknown; however, evidence exists from both scientists and fishermen that sharks do venture into Colombian rivers. Thorson (1972) recorded that the shark, *C. leucas*, was caught 12 km upstream of Leticia, Colombia in 1969 by a local fishermen. Leticia is located on the Amazon 3,480 km upriver from the Atlantic Ocean. The shark was a 2.5 m, 89 kg female. Photographs were taken and the jaws were preserved as evidence. Thorson (1972) also reported a *c.a.* 2.0 m shark landed near Leticia. Reports estimate that 8 to 10 sharks per year are caught near Leticia and sold in local markets (Thorson, 1972). Sharks do not appear to occur in large numbers at any point in the Amazon basin, but they can be looked for occasionally in the Amazon proper as well as its major tributaries any place in the lowlands where the water temperature is suitable and the elevation gradient is moderate (Thorson, 1972). Miles (1947) reported that unidentified sharks (possibly *Carcharhinus* among others) were observed in the Magdalena River and have been observed near the village of Calamar, Bolivar state (10° 15′N, 74° 55′W). These sharks have been known to cause considerable damage to fishing nets at Calmar port on the Magdalena River (Miles, 1947). The sawfish, *Pristis pectinatus* (Latham 1794) (6m *Ls*), have been reported to enter the Magdalena river from the Atlantic sea and have been captured at Puerto Wilches, Santander State (7°535 20′N, 7° 20′E), 600 km from the Atlantic Ocean (Miles, 1947).

CONSERVATION

According to the IUCN, by the end of 2007, almost half (591) of all chondrychthyans had been evaluated at a global scale and 126 species or 21.3% were considered threatened (Dulvy and Forrest, 2010). Two species (3.7%) were considered critically endangered, 29 (4.9%) endangered, and 75 (12.7%) were classified as vulnerable (Dulvy *et al.*, 2008). A further 117 species (18%) were listed as near threatened largely on the basis of the ongoing or increasing degree of potential threats faced by these species (Dulvy and Forrest, 2010). It may or may not be a surprise that there are a large number of species about which little is known—201 species (34%) were listed as data deficient (Dulvy and Forrest, 2010). These reports illustrate the critical status of elasmobranchs worldwide. Conservation measures have been taken in some countries.

Elasmobranch conservation action plans have been adopted in Central and South American countries such as Chile, Costa Rica, Ecuador, Mexico, Uruguay and Brazil. The IUCN (IUCN, 2011) status of the 96 species (26 families) of elasmobranchs reported in Colombia assigns about 24% of these species near threatened, 49% data deficient, 5% least concerned, 2% endangered, 4% critically endangered and 1% not evaluated (Figure 3). Colombia recently developed a National Action Plan for Conservation and Management of Sharks, Rays and Chimeras (Caldas *et al.*, 2010). This Action Plan was in response to FAO´s proposal for developing conservation and management practices.

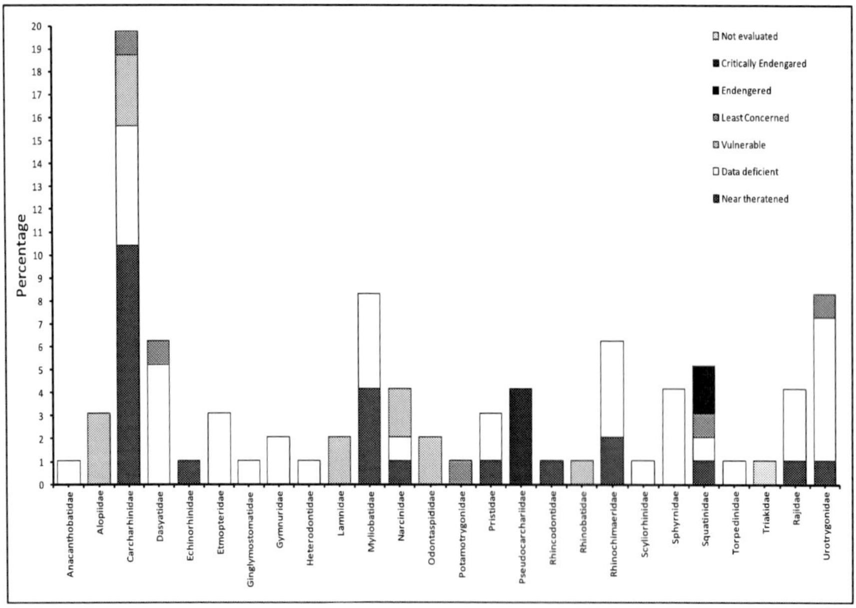

Figure 3.

The objective of this action plan is to guarantee the conservation and sustainable management of sharks, rays and chimeras in marine and continental waters of Colombia that interact with tourist and cultural activities, with national fisheries at the sportsman and commercial level, and among neighboring countries (Caldas *et al.*, 2010). The action plan establishes several components: 1) knowledge diagnosis 2) research (fisheries, bio-ecological, social-economical, and cultural) 3) legislation requirements 4) education and outreach 5) governability, surveillance and control 6) evaluation and monitoring (Caldas *et al.*, 2010). The plan establishes shark, ray and chimera species that have a priority for conservation in the Pacific Ocean (21 shark and 11 ray species) according to IUCN criteria. Shark, ray and chimera conservation is also focused on the Continental Caribbean Ocean (24 sharks and 9 rays) according to the IUCN criteria and 19 shark species and one ray and chimera species for the Caribbean Islands (Caldas *et al.*, 2010). The plan also includes conservation measures for seven species of freshwater stingrays. *P. motoro, P. magdalenae*, *P. orbignyi* and *P. schroederi* have a very high priority conservation action level (Caldas *et al.*, 2010). *Paratrygon aiereba* (Müller and Henle, 1841) and *P. constellata* have a high priority conservation action level. *P. yepezi,* has a mid priority level of action (Caldas *et al.*, 2010). Marine species, *S lewini*, *Ginglymostoma cirratum* (Bonnaterre 1788), I. oxyrinchus, C. falciformis, C. leucas, C. limbatus, *Carcharhinus porosus* (Ranzani 1839) and G. cuvier, require a high priority conservation and management level (Caldas et al., 2010).

Despite the mandate, few laws or regulations have been established for elasmobranchs in Colombia. This is evidenced by only two administrative acts for marine and coastal species: Resolution 1633 of 2007 by Colombia´s Rural Development Institute (Instituto Colombiano de Desarrollo Rural - INCODER) prohibiting shark finning, which is currently under modification, and Resolution 003333, passed in 2008 by the Instituto Colombiano Agropecuario – ICA, which prohibits sports or commercial shark fishing around the San Andrés Archipelago, Providencia and Santa Catalina State (Caldas *et al.*, 2010). Freshwater stingrays are protected by passage of Colombian resolution 3532 in 2007 by INCODER regulating ornamental fish exploitation and marketing (Caldas *et al.*, 2010). Ornamental fish exploitation is the major threat for stingrays in Colombia. Little can be done in places where governmental actions are insipient, negligible or where narco-guerilla groups operate. Potamotrygonid sting rays are valued in the ornamental fish trade. Ornamental fish trade statistics are not reliable because an unknown and possibly large proportion of all exports are made illegally via Colombia from the Amazon region (Rodríguez *et al.*, 2007); however, overexploitation does not seem to be a major concern because operations are on a relatively small scale (Rodríguez *et al.*, 2007). Barletta *et al.* (2010) reported that ornamental fish captures have never been a serious issue in the Orinoco Basin. Today, most ornamental fishes collected in the Orinoco Basin are transshipped to Colombia (Barletta *et al.*, 2010). Despite fish habitat changes and losses, data on fish and fisheries sustainability generated by the Colombian government through the Agriculture and Rural Development Ministry are focused solely on fisheries regulation (*e.g.* closed seasons, minimum capture sizes and gear specifications) and compensatory measures (*e.g.* stocking and aquaculture). More consistent and systemic measures are urgently needed (Barletta *et al.*, 2010).

Conservation efforts based on monitoring shark and ray populations offer great opportunities to access information on their biology. Recently, Mejía-Falla *et al.* (2010a) reported on a monitoring program especially designed for elasmobranch species, such as sharks and rays, and chondricthian species, such as chimeras, for San Andrés and Providence Islands in the Colombian Caribbean Ocean. The program objective is to collect information about the richness, distribution and abundance of chondricthian species. It involves training fishermen, scuba diving instructors and non-instructors on shark and ray identification and data collection. A booklet was developed by the authors for species identification based on data thus far collected. It provides information on ten shark and nine ray species. Based on their results the number of known ray species has increased from six to nine for their study area.

The conservation status of most freshwater and euryhaline elasmobranchs has not been determined. This is probably due to a combination of taxonomic problems and a general lack of information (Martin, 2005). Guitarfishes (Rhinobatidae) and sawfishes (Pristidae) are highly sensitive to fishing pressure as they are large bodied and presumably have a low intrinsic rate of population increase (Dulvy and Forrest, 2010). *Pristis perotteti* (Müller and Henle 1841), has a very high priority conservation action level included in the National Action Plan for Conservation of Elasmobranchs of Colombia (Caldas et al., 2010). It seems likely that successful conservation of freshwater and euryhaline elasmobranchs, even of endemic species with restricted ranges, will require international cooperation (Martin, 2005).

FISHERIES

Worldwide fisheries are the main threat to elasmobranch populations. Fisheries can impact sharks both directly (Walker, 1998; Bonfil 1994; Punt *et al.*, 2005; Duvly and Forrest, 2010) and indirectly by landing enlasmobranchs as bycatch during demersal fish trawling (Graham *et al.*, 2001; Ellis *et al.*, 2005; Shepherd and Myers, 2005; Duvly and Forrest, 2010), crustacean fishing (Stobutzki, *et al.*, 2001; Duvly and Forrest 2010; Straube *et al.*, 2011), pelagic line fishing (Gilman *et al.*, 2008), pelagic trawling (Zeeberg *et al.*, 2006), recreational fishing (Anderson, 2002) or through finning of sharks captured, mainly as bycatch, in pelagic fisheries (Clarke *et al.*, 2006). Few fishing nations report the species composition of their shark catches or landings, obscuring which countries are engaged in pelagic shark fishing and, thus, where management efforts are needed (Camhi *et al.*, 2008).

Inconsistent data are available on elasmobranch fisheries in Colombia. The government poorly monitors elasmobranch fishing. Recorded statistics include broad taxonomic categories that are unreliable for capture estimates (Caldas et al., 2009); however, there are at least 12 species of elasmobranchs that have commercial value and are fished. This contrasts with 22 species of elasmobranchs (14 sharks and eight rays) that are captured as bycatch and six species that are rarely captured (Caldas et al., 2009). Available information about elasmobranch catches has been provided by environmental and fishery institutions that are part of the Colombian government (INCODER, INPA, and the closed INDERENA) and other non-governmental institutions, such as Fundación MAPLELO and Fundación SQUALUS. The Colombian government reports on average about 90 T extracted annually between 1957 and 2004 (Caldas et al., 2009). The values over this period range between about 14 and 304 T (Caldas et al., 2009). Mejía-Falla and Navia (2011b) recently provided evidence about Colombian Pacific shark and ray fisheries based on governmental information (ICA, INCODER between 1975-2011) and other sources (Fundación MALPELO between 2006-2011 and Fundación SQUALUS between 2001-2007). According to INCODER, shark and ray landings from 1985 to 2006 were 25.19% and 3.73%, respectively in Colombia. They provided landing information about industrial and non-industrial shark and ray fisheries from 1975 to 2009 from various ports in the Colombian Pacific Ocean including Buena Ventura (non-industrial shark 961T and ray 93.9T, industrial ray 9.8T and shark 5583.5T), Bahía Solano (non-industrial ray no data recorded, shark 83.5T, industrial ray and shark no data recorded), Guapi (non-industrial ray 10.9T and shark 9.8T, industrial ray and shark no data reported), and Tumaco (non-industrial ray 7.4T and shark 1083.5T, industrial ray 0.8T and shark 343T). Shark and ray landing information provided by governmental institutions (INCODER) is incomplete as information is lacking at some landing ports, e.g. Tumaco, Bahía Solano and Guapi from 2000 to 2005 in the Colombian Pacific Ocean. These data used broad categories failing to identify shark and ray captures to species level, and in some cases the landing ports were unspecified. Other sources of information about shark and ray fisheries have been provided by non-governmental organizations (Mejía-Falla *et al.*, 2011b). Sharks comprised 18,813 kg and rays 6,197.8 kg of the total landings reported by Fundación MALPELO between 2006 and 2009 at Buenaventura in the Colombian Pacific Ocean. Most captured shark species (>900 kg) included *Alopias pelagicus* (Nakamura 1935) (9,765.8 kg), *Alopias* sp. (900.7 kg), *C. falciformis* (1,590.2 kg), *M. lunulatus* (1,266.1 kg), *S. lewini* (4,226.3 kg). Ray species (>900 kg) included *D. longa* (1,731.2 kg) and *Dasyatis dipterura* (Jordan and Gilbert 1880) (1,294.6 kg).

Concern exists about most shark and ray species landed at Bunenaventura as most have not attained sexual maturity (Mejía-Falla *et al.*, 2011b). The Gulf of Tortugas in the Colombian Pacific Ocean has been subject to intense fishing since 1960 (Navia *et al.*, 2012). Although the target of these fisheries is shallow-water shrimp, a large number of fish and invertebrate species are also caught as bycatch (Navia et al., 2012). Shrimp trawler data, indicated that seven shark and 12 ray species were landed between the Gulf of Tortugas and Guapi (data collected in 1995, 2001, 2003, 2004 and 2007) in the Colombian Pacific Ocean (Mejía-Falla and Navia, 2011b). Shark species landings included: *M. lunulatus*, *M. henlei*, *C. porosus*, *Sphyrna corona* (Springer 1940), *S. lewini*, *Sphyrna media* (Springer 1940), and *S. tiburo*. Ray landings included *Narcine entemedor* (Jordan and Starks 1895), *N. leoparda*, *R. leucorhynchus*, *Z. xyster*, *R. velezi*, *Urotrygon aspidura* (Jordan and Gilbert 1882), *U. rogersi*, *Urotrygon* sp., *D. longa*, *Dasyatis* sp., *Gymnura marmorata* (Cooper 1864) and *A. narinari*. Catch per unit effort (CPUE) values (a total of 396 fishing effort hours recorded in 1995, 2001, 2003, 2004 and 2007) were reported for shark and ray species. CPUE for sharks such as *M. lunulatus*, was 0.48 in 1995 (fishing effort hours=76) and reduced to 0.01 in 2007 (fishing effort hours= 110). *C. porosus* CPUE was 0.24 in 1995 (fishing effort hours= 76) and reduced to 0.02 in 2004 (fishing effort hours= 60), *S. tiburo* CPUE was 0.03 in 1995 (fishing effort hours= 76) and slightly reduced to 0.02 in 2007 (fishing effort hours= 110), and *S. lewini* CPUE was 0.04 in 1995 (fishing effort hours= 76) and reduced to 0.02 in 2004 (fishing effort hours= 60) (Mejía-Falla and Navia 2011b). Some ray CPUE values increased, such as in *R. leucorhynchus* from 0.21 in 1995 (fishing effort hours=76) to 0.61 in 2007 (fishing effort hours= 110) and in *U. aspidura* from 0.79 in 1995 (fishing effort hours=76) to 1.32 in 2007 (fishing effort hours= 110). In contrast, *U. rogersi* values decreased from 0.54 in 1995 (fishing effort hours=76) to 0.43 in 2007 (fishing effort hours= 110). They only included the CPUE values for shark and ray species wherein data are reported for most years. They found that shark CPUE values show a decreasing pattern when comparing years. Rays show increasing values and in some cases decreasing CUPE values. Moreover, when comparing all years covered within the study, there are variable CUPE values for both sharks and rays (data not shown). Mejía-Falla and Navia (2011b) concluded that shrimp trawling negatively impacts shark and ray numbers as evidenced by reduced numbers of upper-food web elasmobranchs in heavily trawled areas of their study region in the Colombian Pacific Ocean. Navia *et al.* (2010b) studied the economics and distribution of non-industrial, fishery-related activities at Punta Ardita, Juradó, Piña, Cupica, Bahía Solano and Huina located in the Northern Zone of the Colombian Pacific Ocean. Elasmobranch captures were made at Punta Ardita, Cupica and Bahía Solano. They were represented by six families: Alopiidae, Carcharhinidae, Dasyatidae, Myliobatidae, Sphyrnidae and Triakidae. Navia *et al.* (2008) also studied shark fisheries at six locations in the Colombian Pacific Ocean: Northern Colombian Pacific (Bahía Solano and Ensenada de Urtia), Central Colombian Pacific (Juanchaco and Buenaventura) and Southern Colombian Pacific (Guapi and Tumaco). The study reported fishing gear, economic chains, shark processing and related aspects.

Multiyear information on elasmobranch fisheries in Colombia have been collected by FAO (2011). Shark, ray and chimera landings numbered ≤300 T between 1968 to 1971, 1979, 1995, 2003 and increased to >600 T in 1972 to 1976, 1986 to 1987, and 1989 in the Atlantic and Pacific Oceans (data put together) (Figure 4). Landing reports from the years 2000 to 2009 indicate a decrease in tonnage from 2000 to 2003 (from about 700 T to 10 T), an increase during 2004 (about 500 T) and a progressive increase from 2005 to 2009 (from 400 to 550 T). Mean yearly elasmobranch captures, decreasing from 157 T in the 1960s to 76 T in the last decade, have been recorded for the Colombian Atlantic Ocean (Figures 1 and 4). Similarly, elasmobranch captures decreased from 539 T in the 1980s to 353 T in the last decade in the Pacific (Figure 2 and 4). The data provided by FAO does not include estimates from illegal fish landings that are common in both Atlantic and Pacific Oceans. García *et al.* (2007) reported negative lineal trends of mean biomass with time for rays and sharks during 13 cruises from 1970 to 2001 [1970 (3 cruises), 1988 (4 cruises), 1992 (2 cruises), 1995 (2 cruises), 1996 (1 cruise) and 2001 (1 cruise)] in the upwelling Colombian Caribbean Sea (García *et al.*, 2007). Reported demersal shark landings revealed about 440, 450, 180, and 50 kg km^{-2} in 1979, 1988, 1990s and 2001, respectively. This suggests that although mean shark biomass did decrease with time, individual landing numbers were highly variable (García *et al.*, 2007).

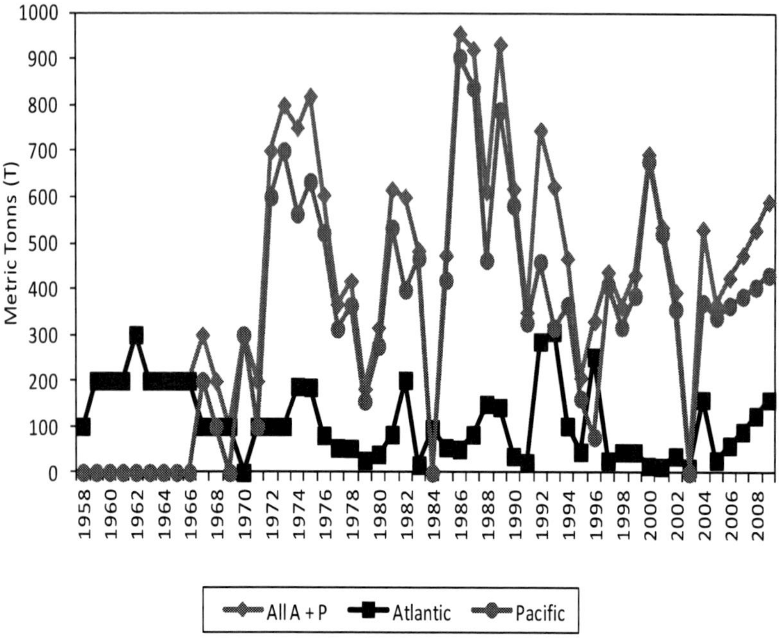

Figure 4.

Reports of ray landings in 1970 were about 6,000 kg km^{-2} and decreased to < 100 kg km^{-2} in 1988, 1990s and 2001. The biomass of rays by year 2001 had not yet returned to 1970 levels and mean biomasses of sharks steadily decreased over the same period (García *et al.*, 2007). Direct elasmobranch fishing occurs in nine zones of the Caribbean Sea such as Isla Fuerte, Las Flores, Ciénaga, Tasajera, Don Jaca, Taganga, Arrecifes, Manaure and Cabo de la Vela among others (Caldas *et al.*, 2009).

Elasmobranchs are captured incidentally as bycatch by both sport and commercial fisheries in Colombia (Grijalba-Bendeck *et al.*, 2008). This is evidenced by total elasmobranch discards from bycatch (75% discards or 1,445 T) during 2000 in the Atlantic Ocean (Manjarrés, 2003). This seems to be the main cause of population decreases. Attrition resulting from bycatch and shrimp fisheries (sharks being caught accidently in shrimp fishing nets) clearly impact shark populations in Colombia (Moreno *et al.*, 2008). The endemic species *Mustelus minicanis* (Heemstra 1997) that has a distribution from Cabo de la Vela in Colombia (Atlantic Ocean) to Rio Caribe (Venezuela) could be endangered as a result of bycatch and shrimp fisheries (Moreno *et al.*, 2008). Low species abundance e.g. *Z. xyster* could indicate overfishing of this species in the central zone off the Colombian Pacific Ocean (Mejía-Falla *et al.*, 2006).

FRESHWATER ELASMOBRANCHS

At least 171 species of elasmobranch, representing 68 genera and 34 families, are recorded from fresh or brackish estuarine waters (Martin, 2005). Among these, the exclusively freshwater stingray family Potamotrygonidae ranges throughout the major Atlantic drainages of South America (Rosa, 1985; Rosa *et al.*, 2010).

Obligate freshwater elasmobranchs are dominated by myliobatoid stingrays, of which two-thirds are potamotrygonids endemic to Atlantic drainages of South America (Martin, 2005). *Potamotrygon* occur in all South American countries with the exception of Chile. They are found in the drainages of the Atrato and Magdalena Rivers (Colombia), the Orinoco and Maracaibo Rivers (Venezuela), the coastal rivers of the Guianas, Surinam, and Amapá State in Brazil and the Amazon River and its tributaries in Bolivia, Brazil, Colombia, Ecuador, Peru and Venezuela.

Other rivers where *Potamotrygon* occurs are at the Tocantins River, coastal rivers of Maranhao and Piaui States in Brazil, including Parnaiba River, and the Parana-Paraguay basin (Argentina, Brazil, Paraguay, and Uruguay), including Rio Uruguay (Rosa *et al.*, 2010). Potamotrygonids have been reported from various Colombian streams, rivers and floodplains.

Scant research has been conducted on Colombian elasmobranchs inhabiting freshwaters, and little is known about the potamotrygonid stingrays there. Most information has been provided by checklists reporting their distribution in Colombian rivers such as the Amazon, Magdalena, Cauca, Orinoco, Meta, Putumayo and some of their tributaries (Mojica-C, 1999; Mojica *et al.*, 2005, 2006; Lasso *et al.*, 2004, 2009; Villa-Navarro *et al.*, 2006; Maldonado-Ocampo *et al.*, 2006, 2008). *P. magdalenae* is an endemic species found in the Magdalena River Basin (Teshima and Takeshita, 1992; Villa-Navarro *et al.*, 2006; Mojica *et al.*, 2006).

This species has been reported in the Catatumbo, Magdalena, Cauca and San Jorge Rivers (Mojica-C, 1999). *P. hystrix* has been found in the Putumayo river (Mojica-C, 1999), Mesay river, streams and lagoons (Caquetá, Colombian Amazon) (Bejarano *et al.*, 2006). *P. motoro* is indigenous to the Meta river (Mojica-C, 1999; Lasso *et al.*, 2004), Cauca (Lasso *et al.*, 2004), Orinoco (Lasso *et al.*, 2004), Tomo river, Orinoco (Maldonado-Ocampo *et al.*,

2006) and Inírida rivers (Lasso *et al.*, 2009). *P. yepezi* is indigenous to the Catatumbo river (Mojica-C, 1999). *Paratygon aiereba* has been found in the Tomo river, a tributary of the Orinoco (Maldonado-Ocampo *et al.*, 2006) and the Orinoco river (Lasso *et al.*, 2009). *P. orbignyi*, was reported in the Tomo, (Maldonado-Ocampo *et al.*, 2006), Meta (Lasso *et al.*, 2004), Orinoco and Guaviare Rivers (Lasso *et al.*, 2009). *P. schroederi* occurs in the Orinoco River (Lasso *et al.*, 2004; Lasso *et al.*, 2009).

The life history of freshwater stingrays in Colombia is virtually unknown. Biologists have often encountered difficulties in identifying potamotrygonid stingrays, and the same has happened with researchers working on medical aspects of the group, not to mention fish traders and breeders, usually less concerned with taxonomy (Rosa *et al.*, 2010).

A few studies have focused on the biology of Potamotrygonid species. Teshima and Takeshita (1992) examined the reproduction of *P. magdalenae* found in the Magdalena River. Teshima and Hara (1983) examined the epithelial cells of the intestines of *P. magdalenae* from the Magdalena River. Columnar cells lining the internal surface of the intestine in a single layer form microvilli that are densely arranged on the free surface. Intermicrovillous invaginations, numerous vesicles beneath the microvilli and granules of varying sizes apparently aid in nutrient absorption. Brooks and Thorson (1976) reported two new parasites (tetraphyllidean cestodes) infesting the freshwater stingray, *P. magdalenae* from Colombia.

Aside from these scientific studies, information comes from the ornamental fish trade, wherein potamotrygonids are highly valued (Mancera-Rodríguez, and Álvarez-León, 2008; Mejía-Falla *et al.*, 2009; Prada-Pedreros *et al.*, 2009). Sting rays are not valued for human consumption and when fished are often discarded as bycatch. *P. motoro* and *Potramotrygon* sp., are discarded as bycatch in the Meta and Orinoco Rivers on the border between Colombia and Venezuela (Camargo-Guerra *et al.*, 2001). Mejía-Falla *et al.* (2009) reported that *P. magdaleane* is sometimes used for human consumption; however, Galvis *et al.* (1997) reported that it is not important as such. Juvenile *P. motoro* rays are part of the ornamental fish trade in Colombia (Maldonado-Ocampo, 2000). Galvis *et al.* (1997) reported that *P. yepezi* is very frequent in fish captures on the Catatumbo River, Colombia; however there is evidence that this species is not so common (Mejía-Falla *et al.*, 2009). Fishermen avoid injuries from spine punctures by killing *P. yepezi* in Catatumbo as these rays have caused serious injuries to fisherment (Galvis *et al.*, (1997). Sting ray export to other counties has been poorly monitored by Colombia's Governmental organizations and only incomplete internal reports exist. Another problem surrounding ray exportation is the number of illegally shipped specimens. It is impossible to know the actual exported number.

Currently, there is no strong enforcement by the government apart from action plans that most of the time remain only on paper and are never emplimented. Freshwater sting ray exports have been reported by the Instituto Nacionl de Pesca y Agriculture (Colombia's fisheries institution) between 1993-2000 (no data reported for 1994 and 1997) to be 22,343 individuals, and INCODER, (replaced former Instituto Nacionl de Pesca y Agricultura) reported an average of 30,500 individuals exported in 1994 and 2006. Important rays reported by the Colombian ornamental trade include: *P. motoro*, *P. magdalenae*, and *P. shroederi* among other species (Mejía-Falla *et al.*, 2009). *P. magdalenae* is highly valued as an ornamental fish, and exports represent 60 to 70% of all ray species exported, mainly from the lower Magdalena River at Cienaga Grade of Santa Marta near the Colombian Craibbean Ocean (Dahl, 1971; Araújo *et al*, 2004; Perdomo, 2005). Nearly 550 individuals of *P. motoro* were captured by the ornamental trade sector between May, 1998 and April 1999 at Inirida, Guainía State, Colombia (Ramírez-Gil and Ajiaco-Martinez, 2001). Mejía-Falla *et al.* (2010b) described the biological, fishing, social and economic impacts of ray fishing at Puerto Carreño, Colombian Orinoco. Their study was based on fisherman surveys that included their activities, place and timing for ray fishing, collectors, prices by collectors, exporters and pricing, ray management under captive conditions and payments. The study concluded that the ray market is based mainly on demand from other countries and, is therefore difficult to predict the amount of specimens exported to other countries.

CONCLUSION

Little information is available on elasmobrach biology when compared to the total number of reported elasmobranch species (124). Recently (2000-present), there has been an increase in the number of species investigated; however this number is about 10% of the total number of species recorded in Colombia. Freshwater sting ray species may need taxonomic revisions. Feeding ecology studies are based mainly on stomach content analyses and food web interactions on benthic and, in some cases, demersal elasmobranchs. These studies have mainly focused on rays and less on sharks. Future studies should focus on ecomorphological aspects of elasmobranchs relating to morphology, diet and habitat in order to better understand their ecological niche. Special care should be taken when conducting diet analyses due the number of individuals killed per study because populations could be impacted. Although research provides important insights into elasmobranch biology and ecology, there is a growing concern that conducting research on data-deficient or endangered species using destructive methods could impact the survival of those species.

Feeding ecology studies have been conducted on a few freshwater sting ray species, but there is a broad range of species that have no ecological information available. Among these are *Paratigon aereiba*, *P. constellata*, *Potamotrygon motoro*, *Potamotrygon obrignyi*, *Potamotrygon shroederi* and *Plesiotrygon iwamae*. Future research should focus on feeding ecology and reproductive aspects in addition to the general biology of the above cited rays. Reproductive aspects are better documented on elasmobranchs in the Caribbean Ocean than in the Pacific; however, more studies are required to further understand their reproductive cycles. The only shark migration study up to date reports on the endangered *S. lewini* off the Colombian Pacific Ocean; however, more research is required on the topic. Sharks venture into Colombian rivers. Particularly the genus *Carcharhinus* and mainly *C. leucas*, has been reported to venture into Colombian rivers such as the Amazon, near Leticia and in northern rivers. Their migratory range could include other rivers. Elasmobranch fishery statistics are currently inconsistent as most governmental institutions have unreliable data sets that are both general and report broad taxonomic categories. Based on this, the Colombian government should develop action programs to consolidate a reliable and consistent, system-based methodology to manage fishery, economic, social and related information. Special care should be taken in regulation of elasmobranch captures below the minimum reproductive size because this could eventually lead to population decline. More research is required establish size limits. Governmental efforts should focus more on regulatory activities and less on paper.

Fishery conservation measures, such as elasmobranch species capture bans and special bycatch-evasive maneuvers should be developed in order to minimize capture impacts. Conservation measures should be taken to regulate and control the ornamental freshwater ray's illegal trade that is based on demand from developed countries. Little has been done in regions of Colombia where governmental presence is negligible or almost negligible where narco-guerilla groups operate. Conservation efforts should focus more on education.

Educating future generations that are directly involved with elasmobranch fisheries e.g. fishermen families, ornamental fish collectors and distributors as well as the general public should be considered a basis for elasmobranch conservation. More research is required on all aspects of elasmobranch biology. Furthermore, future investigations should focus on physiological aspects, ecomorphology, population estimates, genetics and fishery impact assessment. It is hoped that Colombian scientists can broaden their research interests to develop a better understanding of the biology of indigenous elasmobranchs. This chapter was written in an effort to motivate creative young scientists to research the interesting world of Colombian elasmobranchs.

ACKNOWLEDGMENT

Special thanks to L. Marien for her valuable support in this investigation.

REFERENCES

Acero, A., Canter-Rios, D. and Polanco, A. (2007). Identidad problemática del tiburón martillo ojichico (Carcharhiniformes: Sphyrna) existente en Colombia. *Boletín de Investigaciones Marinas y Costeras-INVEMAR*, 36, 321-326.

Acero, A., Grijalba-Bendeck, M., Moreno, F., Acevedo, K. and González, E. (2008). Histología comparada del sistema reproductivo de cuatro especies de rayas caribeñas (Chondrichthyes: Batoidea). *Revista de la Academia Colombiana de Ciencias*, 32, 571-586.

Acevedo, K., Bohórquez-Herrera, J., Moreno, C., Molina, E., Grijalba-Bendeck, M. and Goméz-Canchong P. (2007). Aspectos reproductivos de tres Rajiformes capturados incidentalmente en la pesquería camaronera de arrastre en el mar Caribe de Colombia. Annual *Gulf and Caribbean Fisheries Institute,* 58, 229-236.

Adams, P. B. (1980). Life history patterns in marine fishes and their consequences for fisheries management. Fishery Bulletin 78, 1-12.

Aires-da-Silva, A., Ferreira, R. L. and Pereira, J. G. (2008). Case study: Blue shark catch-rate patterns from the Portuguese swordfish longline fishery in the Azores. In Sharks of the Open Ocean: Biology, Fisheries and

Conservation. (Camhi, M. D. and Pikitch, E.K., eds), pp 230-235. Oxford, UK: Blackwell.

Anderson, R. C. (2002). Elasmobranchs as a recreational resource. In Elasmobranch Biodiversity, Conservation and Management. (Fowler, S. L., Reid, T. M. and Dipper F.A., eds). Proceedings of the International Seminar and Workshop. Sabah, East Malaysia. July 1997, pp. 46–51. Occasional Paper of the IUCN Species Survival Commission No 25. IUCN The World Conservation Union: Gland, Switzerland.

Araújo, M. L. G., Charvet-Almeida, P., Almeida, M. P. and Pereira, H. (2004). Freshwater stingrays (Potamotrygonidae): status, conservation and management challenges. *Information Document AC20*, 8, 1-6.

Argumentado, E. 2008. Experiencias regionales sobre manejo de rayas del género Potamotrygon en caitiverio y propuesta de aprovechamiento sostenimble In: Payán, L.F., Muñóz-Osorio, L. A., Ramírez-Luna, V. and Mejía-Ladino, L. M. (Eds). Libro de resúmenes I. Encuentro colombiano sobre condrictios. Funadacion SQUALUS. p 52.

Ballantyne, J. S. and Robinson, J. W. (2010). Freshwater elasmobranchs: a review of their physiology and biochemistry. *Journal of Comparative Physiology B 180,* 475–493.

Barletta, M., Jaureguizar, A. J., Baigun, C., Fontoura, N. F., Agostinho, A. A., Almeida-Val, V. M. F., Val, A. L., Torres, R. A., Jimenes-Segura, L. F., Giarrizzo, T., Fabré, N. N., Batista, V. S., Lasso, C., Taphorn, D. C., Costa, M. F., Chaves, P. T., Vieira J. P. and Corrêa. M. F. M. (2010). Fish and aquatic habitat conservation in South America: a continental overview with emphasis on neotropical systems. *Journal of Fish Biology*, 76, 2118–2176.

Bejarano, I., Blanco, M. P. and Mojica, J. I. (2006). La comunidad íctica del Río Mesay durante el periodo de aguas altas (Caquetá, Amazonía Colombiana). *Caldasia*, 28, 359-370.

Bessudo, S., Soler, G. A., Klimley, A. P., Ketchum, J.T., Hearn, A. and Arauz, R. (2011). Residency of the scalloped hammerhead shark (S. lewini) at Malpelo Island and evidence of migration to other islands in the Eastern Tropical Pacific. *Environmental Biology of Fishes*, 91, 165-176.

Bigelow, H. B. and Schroeder, W. C. (1953). Sawfishes, guitarfishes, skates and rays; chimaeroids. In Fishes of the Western North Atlantic. Vol. 1, Part 2, New Haven, CT; Sears Foundation for Marine Research.

Bone, Q. and Moore, R. H. (2008). Biology of fishes. New York: Taylor and Francis Group.

Bonfil, R. (1994). Overview of world elasmobranch fisheries. Rome: FAO Fisheries Technical Paper 341.

Brooks, D. R. and Thorson T. B. (1976). Two tetraphyllidean cestodes from the freshwater stingray Potamotrygon magdalenae Dumeril 1852 (Chondrichthyes: Potamotrygonidae) from Colombia. *The Journal of Parasitology,* 62, 943-947.

Caillet, G. M., Love, M. S. and Ebeling, A. W. (1996). Fishes: A Field and Laboratory Manual on their Structure, Identification and Natural History. Prospect Heights, Ill; Waveland Press.

Caldas, J. P., Díaz-Trujillo, E. M., and Duarte, L. O. (2009). Revisión histórica de la pesca de tiburones y rayas en el mar Caribe continental de Colombia. In Avances en el Conocimiento de Tiburones, Rayas y Quimeras de Colombia. (Puentes, V. A., Navia, F., Mejía-Falla, P. A., Caldas, J. P., Diazgranados, M. C., y Zapata, L. A., eds), pp 99-129. Fundación Squalus, Ministerios de Ambiente, Vivienda y Desarrollo Territorial, Instituto Colombiano Agropecuario, Colciencias, Conservación Internacional, WWF.

Caldas, J.P., Castro-González, E., Puentes, V., Rueda, M., Lasso, C., Duarte, L.O., Grijalba-Bendeck, M., Gómez, F., Navia, A.F., Mejía-Falla, P.A., Bessudo, S., Diazgranados, M.C. and Zapata Padilla, L.A. (2010). Plan de Acción Nacional para la Conservación y Manejo de Tiburones, Rayas y Quimeras de Colombia (PAN-Tiburones Colombia). Instituto Colombiano Agropecuario, Secretaria Agricultura y Pesca San Andrés Isla, Ministerio de Ambiente, Vivienda y Desarrollo Territorial, Instituto de Investigaciones Marinas y Costeras, Instituto Alexander Von Humboldt, Universidad del Magdalena, Universidad Jorge Tadeo Lozano, Pontificia Universidad Javeriana, Fundación SQUALUS, Fundación Malpelo y otros Ecosistemas Marinos, Conservación Internacional, WWF Colombia. Editorial Produmedios, Bogotá.

Camargo-Guerra, T., Ajuaco-Martínes, R., Beltrán-Galeano, I. and Ramírez-Gil, H. (2001). Selectividad y eficiencia de redes de enmalle en los Ríos Meta y Orinoco, frontera Colombo-Venezolana In: La pesca en la baja Orinoquía colombiana: una visión integral. (Ramírez-Gil, H. and Ajuaco-Martínes, R., eds), pp. 57-77. Bogotá, Instituto Nacional de Pesca y Agricultura, INPA.

Camhi, M. D., Lauck, E., Pikitch E. K., and Babcock. E. A. (2008). A global overview of commercial fisheries for open ocean sharks. In Sharks of the Open Ocean: Biology, Fisheries and Conservation. (Camhi, M. D. and Pikitch, E.K., eds), pp 166-192. Oxford, UK: Blackwell.

Casey, J. G. and Kohler, N. E. (1992). Tagging studies on the shortfin mako (Isurus oxyrinchus) in the western North Atlantic. In Sharks: Biology and Fisheries. (Pepperell, J. G., ed), *Australian Journal of Marine a Freshwater Research 43 (special volume),* 45–60.

Castro-González, E. R. and Ballesteros-Galvis, C. A. (2009). Estado del conocimiento sobre tiburones, rayas y quimeras en el Archipiélago de San Andrés, Providencia y Santa Catalina, Caribe insular colombiano. In Avances en el Conocimiento de Tiburones, Rayas y Quimeras de Colombia. (Puentes, V. A., Navia, F., Mejía-Falla, P. A., Caldas, J. P., Diazgranados, M. C., y Zapata, L. A., eds), pp 11-38. Fundación Squalus, Ministerios de Ambiente, Vivienda y Desarrollo Territorial, Instituto Colombiano Agropecuario, Colciencias, Conservación Internacional, WWF.

Charvet-Almeida, P., Araujo, M.L.G., Rosa, R. S. and Rincon, G. (2002). Neotropical freshwater stingrays: diversity and conservation status. *Shark News*, 14, 1–2.

Charvet-Almeida, P, Araujo, M. L. G. and Almeida, M. P. (2005). Reproductive aspects of freshwater stingrays (Chondrichthyes: Potamotrygonidae) in the Brazilian Amazon Basin. *Journal of the Northwest Atlantic Fishery Science*, 35, 165–171.

Clarke, S. C., McAllister, M. K., Milner-Gulland, E. J., Kirkwood, G. P., Michielsens, C. G. J., Agnew, D. J., Piktich, E. K., Nakano, H. and Shivji, M. S. (2006). Global estimates of shark catches using trade records from commercial markets. *Ecology Letters*, 9, 1115–1126.

Compagno, L. J. V. (1990). Alternate life history styles of cartilaginous fishes in time and space. *Environmental Biology of Fishes*, 28, 33-75.

Compagno, L. J. V. and Cook, S. F. (1995). The exploitation and conservation of freshwater elasmobranchs: status of the taxa and prospects for the future. *Journal of Aquariculture and Aquatic Sciences*, 7, 62-90.

Dahl, G. 1971. Los peces del Norte de Colombia. INDERENA. Bogotá.

de Carvalho, M. R. and Lovejoy, N. R. (2011). Morphology and phylogenetic relationships of a remarkable new genus and two new species of Neotropical freshwater stingrays from the Amazon basin (Chondrichthyes: Potamotrygonidae). *Zootaxa* 2776, 13–48.

Domeier, M. L. and Nasby-Lucas, N. (2008). Migration patterns of white sharks Carcharodon carcharias tagged at Guadalupe Island, Mexico, and identification of an eastern Pacific shared offshore foraging area. *Marine Ecology Progress Series*, 370, 221–237.

Dulvy, N. K., Baum, J. K., Clarke, S., Compagno, L.V. J., Cortes, E., Domingo, A., Fordham, S., Fowler, S., Francis, M. P., Gibson, C., Martínez, J., Musick, J. A., Soldo, A., Stevens, J. D. and Valenti, S. (2008). You can swim but you can't hide: the global status and conservation of oceanic pelagic sharks. *Aquatic Conservation*, 18, 459–482.

Dulvy, N. K. and Forrest. R. E. (2010). Life histories, population dynamics, and extinction risks in chondrichthyans. In Sharks and their Relatives II: Biodiversity, Adaptive Physiology, and Conservation (Carrier, J. C., Musick, J. A. and Heithaus, M. R., eds), pp 639-679. Boca Ratón, FL; CRC Press, Taylor and Francis Group.

Ellis, J. R., Dulvy, N. K., Jennings, S., Parker-Humphreys, M. and Rogers, S. I. (2005). Assessing the status of demersal elasmobranchs in UK waters: a review. *Journal of the Marine Biological Association of the United Kingdom*, 85, 1025–1047.

Ferretti, F., Myers, R. A., Serena, F. and Lotze, H. K. (2008). Loss of large predatory sharks from the Mediterranean Sea. *Conservation Biology*, 22, 952–964.

Galvis, G., Mojica, J.I. and Camargo, M. 1997. Peces del Catatumbo. Asociación Cravo Norte. Bogotá.

García, C. B., Duarte, L. O., Altamar, J. and Manjarrés, L. M. (2007). Demersal fish density in the upwelling ecosystem off Colombia, Caribbean Sea: Historic outlook. *Fisheries Research*, 85, 68–73.

Garman, S. (1877). On the pelvis and external sexual organs of selachians, with special reference to the new genera Potamotrygon and Disceus. *Proceedings of the Boston Society of Natural History*, 19, 197–215.

Garman, S. (1913). The Plagiostomia (sharks, skates and rays). *Memoirs of the Museum of Comparative Zoology*, 36, 1–515.

Gilman, E., Clarke, S., Brothers, N., Alfaro-Shigueto, Mandelman, J., Mangelf, J., Petersenh, S., Piovanoi, S., Thomson, N., Dalzellk, P., Donosol, M., Gorenh, M. and Werner, T. (2008). Shark interactions in pelagic longline fisheries. *Marine Policy*, 32, 1–18.

Goldman, K. J. and Musick, J. A., (2008). The biology and ecology of the salmon shark, Lamna ditropis. In Sharks of the Open Ocean: Biology, Fisheries and Conservation. (Camhi, M. D. and Pikitch, E.K., eds), pp 95-104. Oxford, UK: Blackwell.

Gómez, G. O., Zapata, L. A., Franke, R. and Ramos, G. E. (2003). Food habits of Mustelus Lunulatus and M. henlei (Pisces: Triakidae) colected in

Parque Nacional Natural Gorgona, Colombian Pacific. Boletin de Investigaciones Marinas y Costeras-INVEMAR 32, 219-229.

Graham, K. J., Andrew, N. L. and Hodgson, K. E. (2001). Changes in relative abundance of sharks and rays on Australian South East Fishery trawl grounds after twenty years of fishing. *Marine and Freshwater Research*, 52, 549–561.

Grijalba-Bendeck, M., Acero, P. A. and González, E. (2008). Biología reproductiva de Rhinobatos percellens (Walbaum, 1792) (Batoidea: Rajiformes) en el caribe colombiano. *Revista de Biología Marina y Oceanografía,* 43, 469-481.

Grijalba-Bendeck, M. and Acevedo, K. (2009). Mitsukurina owstoni Jordan, 1898 (Condrichtyes: Mitsukurinidae) primer registro para el Caribe colombiano. *Boletín de Investigaciones Marinas y Costeras INVEMAR*, 38, 21-215.

Grijalba-Bendeck, M., Acero-Pizarro, A., Díaz-Trujillo, E. M. and Gómez. F. (2009). Estado del conocimiento de los peces cartilaginosos del Caribe continental de Colombia. In Avances en el Conocimiento de Tiburones, Rayas and Quimeras de Colombia. (Puentes, V., Navia, A. F., Mejía-Falla, P. A., Caldas, J. P., Diazgranados, M. C. and Zapata, L. A., eds), pp. 39 - 94. Fundación SQUALUS, Ministerio de Ambiente Vivienda y Desarrollo Territorial, Instituto Colombiano Agropecuario, COLCIENCIAS, Conservación Internacional.

Heithaus, M. R., Burkholder, D., Hueter, R. E., Heithaus, L. I., Pratt, H. W. Jr. and Carrier, J. C. (2007). Spatial and temporal variation in shark communities of the lower Florida Keys and evidence for historical population declines. *Canadian Journal of Fisheries Aquatic Sciences*, 64, 1302–1313.

Heithaus, M. R., Frid, A., Vaudo, J. J., Worm, B. and Wirsing, A. J. (2010). Unraveling the ecological importance of elasmobranchs. In Sharks and their Relatives II: Biodiversity, Adaptive Physiology, and Conservation, (Carrier, J. C., Musick, J. A. and Heithaus, M. R., eds). pp 611-638. Boca Ratón, FL; CRC Press, Taylor and Francis Group.

Helfman, G. S., Collette, B. B., Facey, D. E. and Bowen, B. W. (2009). The Diversity of Fishes. West Sussex. UK: Wiley-Blackwell.

INPA, 1997. Boletín Estadístico Pesquero 1996. Santafé de Bogotá D.C. Colombia.

Hleap, J.S., Lozano, R. A. and Navia, A. F. (2010). Informe técnico sobre elasmobranquios en el PNN Gorgona. Expedición científica 2009. Documento técnico Fundación SQUALUS No FS0110. 43 pp.

Hleap, J.S., Mejía-Falla P.A. and Cárdenas, H. (2012). Relaciones morfométricas de la raya redonda *Urotrygon rogersi*: implicaciones cuantitativas bajo modelos lineales. *Revista de Biología Marina y Oceanografía*. 47, 35-50.

Kohler, N. E., Casey, J. G. and Turner, P. A. (1998). NMFS Cooperative shark tagging program, 1962–93: An atlas of shark tag and recapture data. *Marine and Fisheries Review*, 60, 1-87.

Kohler, N. E. and Turner, P. A. (2008). Stock structure of the blue shark, Prionace glauca, in the North Atlantic Ocean based on tagging data. In Sharks of the Open Ocean: Biology, Fisheries and Conservation. (Camhi, M. D. and Pikitch, E.K., eds). pp 339–350. Oxford, UK: Blackwell.

Lasso C. A., Mojica, J. I., Usma, J. S., Maldonado-Ocampo, J. A., DoNascimento, C., Taphorn, D. C., Provenzano, F., Lasso-Alcalá, O. M., Galvis, G., Vásquez, L., Lugo, M., Machado-Allison, A., Royero, R., Suárez, C. and Ortega-Lara, A. (2004). Peces de la cuenca del río Orinoco. Parte I: lista de especies y distribución por subcuencas. *Biota Colombiana*, 5, 95-158.

Lasso, C. A., Oviedo, U., Villa, J. S., Sierra-Quintero, F., Ortega-Lara, M. T., Mesa, A., Patiño, L. M., Lasso-Alcalá, M. A., Morales-Betancourt, O. M., González-Oropesa, M. A., Quiceno, K., Ferrer, M. P., Suárez, A. and Freddy, C. (2009). Peces de la Estrella Fluvial Inírida: ríos Guaviare, Inírida, Atabapo y Orinoco (Orinoquia Colombiana). *Biota Colombiana*, 10, 89-122.

Lasso C. A., Acosta A. and Agudelo, E. (2010). Primer registro de la raya látigo o antena Plesiotrygon iwamae Rosa, Castello y Thorson 1987 (Chondrichthyes: Potamotrygonidae) para Colombia. *Biota Colombiana*, 11, 173-175.

López, J., Mejía-Falla, P. A. and Navia, A. F. (2009). Aspectos biológicos de la raya látigo *Dasyatis longa* (Pisces: Dasyatidae) de la zona central del Pacifico colombiano. Documento técnico Fundación SQUALUS No FS0109. 39 pp.

López-García, J., Navia, A. F., Mejía-Falla, P.A. and Rubio, E. A. (2012). Feeding habits of Dasyatis longa (Elasmobranchii: Myliobatiformes): sexual, temporal and ontogenetic effects. *Journal of Fish Biology*, 80, 1563-1579.

Lucas, M. C., Baras, E., Thom, T. J., Duncan, A. and Slavík, O. (2001). Migration of Freshwater Fishes. London, UK: Blackwell.

Lutton, B. V., St. George, J. M., C. R., Fileti, L. A. and Callard, I. P. (2005). The elasmobranch ovary. In Reproductive Biology and Phylogeny of

Chondrichthyes: Sharks, Batoids, and Chimaeras, Vol 3 (Hamlett, W. C., ed), pp. 237-281. Playmouth, UK: Science Publishers, Inc.

Maldonado-Ocampo, J. A. (2000). Peces de Puerto Carreño: lista ilustrada. Fundación Omacha, Bogotá. 87 p.

Maldonado-Ocampo, J.A., Lugo, M., Bogotá-Gregory, J. D., Lasso, C., Vásquez, L., Usma, J. S., Taphorn, D. and Provenzano, F. (2006). Peces del río Tomo, cuenca del Orinoco, Colombia. *Biota colombiana*, 7, 113-127.

Maldonado-Ocampo, J. A., Vari, R. P. and Usma, J.S. (2008). Checklist of the freshwater fishes from Colombia. *Biota Colombiana*, 9, 143-237.

Mancera-Rodríguez, N. J. and Álvarez-León, R. (2008). Comercio de peces ornamentales en Colombia. *Acta Biológica Colombiana*, 13, 23-52.

Manjarrés, L. (2003). Valoración biológico pesquera y ecológica de la pesca industrial y arrastre camaronero e impacto de la introducción de dispositivos reductores de fauna acompañante, en el mar Caribe colombiano. Santa Marta Magdalena, Colombia.

Martin, R. A. (2005). Conservation of freshwater and euryhaline elasmobranchs: a review. *Journal of the Marine Biological Association of the United Kingdom,* 85, 1049-1073.

Miles, C. (*1947*). Los *Peces* del Río Magdalena. Ministerio de la Economía Nacional, Sección de Piscicultura, Pesca y Caza, Bogotá, Colombia.

Mejía-Falla, P. A. and Navia, A. F. (2006). Dinámica poblacional y uso de hábitat y del tiburón aletiblanco (*Triaenodon obesus*), del tiburón ballena (*Rhincodon typus*) y de la mantaraya (*Manta birostris*), en el Parque Nacional Natural Gorgona, Pacifico colombiano. Documento Técnico No FS0306 Fundación SQUALUS. 27 pp.

Mejía-Falla, P. A., Navia, A. F. and Giraldo, A. (2006). Notas biológicas de la raya ocelada Zapteryx xyster Jordan and Evermann, 1896 (Chondrichthyes: Rhinobatidae) en la zona central de pesca del Pacífico colombiano. *Investigaciones Marinas, Valparaíso*, 34, 181-185.

Mejía-Falla, R A., Navia, A.F., Mejia-Ladino, L.M., Acero A., and Rubio, E. (2007). Tiburones y rayas de Colombia (Pisces: Elasmobranchii). Lista revisada, actualizada y comentada. *Boletín de Investigaciones Marinas y Costeras-INVEMAR*, 34, 128-169.

Mejía-Falla, P. A. and Navia, A. F. (2009). New records of Urobatis tumbesensis (Chirichigno and McEachran, 1979) in the Tropical Eastern Pacific. *Pan-American Journal of Aquatic Sciences*, 4, 255-258.

Mejía-Falla, P. A., Ramírez, A., Usma, S., Muñoz, L. A., Maldonado-Ocampo, J., Sanabria A. I., and Alonso, J. C. (2009). Estado del conocimiento de las rayas dulceacuícolas de Colombia. In Avances en el Conocimiento de Tiburones, Rayas and Quimeras de Colombia. (Puentes, V., Navia, A. F., Mejía-Falla, P. A., Caldas, J. P., Diazgranados, M. C. and Zapata, L. A., eds), pp. 197-245. Fundación SQUALUS, Ministerio de Ambiente Vivienda y Desarrollo Territorial, Instituto Colombiano Agropecuario, COLCIENCIAS, Conservación Internacional.

Mejía-Falla, P.A., Navia, A. F., López, J. and Lozano, R. (2010a). Programa de monitoreo de tiburones, rayas y quimeras del Archipiélago de San Andrés, Providencia y Santa Catalina-PNAT SEAFLOWER-. Documento técnico Fundación SQUALUS No FS0510. 20 pp.

Mejía-Falla, P. A., Muñoz, L. A., Ramírez-Luna V. and Navia. A. F. (2010b). Caracterización biológica-pesquera y socioeconómica de la pesca de rayas de agua dulce en Puerto Carreño, Orinoquía Colombiana. Documento técnico Fundación SQUALUS No FS0410. 16 pp.

Mejía-Falla, P. A. and Navia, A. F. (2011a). Relationship between body size and geographic range size of elasmobranchs from the Tropical Eastern Pacific: An initial approximation for their conservation. *Ciencias Marinas,* 37, 305-321.

Mejía-Falla, P. A. and Navia, A. F. (2011b). Estadísticas pesqueras de tiburones y rayas en el Pacífico colombiano. Documento técnico Fundación SQUALUS No FS0111. 70 pp.

Mejía-Falla, P. A., Navia, A. F. and Muñoz, L. A. (2011). Morphological abnormality in embryos of Urotrygon rogersi (Myliobatiformes: Urotrygonidae) in the Pacific Coast of Colombia. *Latin American Journal of Aquatic Research,* 39, 184-188.

Mejía-Falla, P. A., Navia, A. F. and Cortés, E. (2012). Reproductive variables of Urotrygon rogersi (Batoidea: Urotrygonidae): a species with a triannual reproductive cycle in the Eastern Tropical Pacific Ocean. *Journal of Fish Biology*, 80, 1246- 1266.

Mojica-C, J. I. (1999). Lista preliminar de las especies de peces dulceacuícolas de Colombia. *Revista de la Academia Colombiana de Ciencias Físicas y Naturales,* 23(Suplemento especial), 547-566.

Mojica, J. I., Galvis, G., Arbeláez, F., Santos, M., Vejarano, S., Prieto-Piraquive, E., Arce, M., Sánchez-Duarte, P., Castellanos, C., Gutiérrez, A., Duque, S. R., Lobón-Cerviá, J. and Granado-Lorencio, C. (2005). Peces de la cuenca del río Amazonas en Colombia: Región de Leticia. *Biota Colombiana*, 6, 191-210.

Mojica, J. I., Galvis, G., Sánchez-Duarte, P., Castellanos, C. and Villa-Navarro, F. A. (2006). Peces del valle medio del río Magdalena, Colombia. *Biota Colombiana*, 7, 23 – 38.

Moreno, F., Acevedo, U. K. and Paramo, J. (2008). Aportes a la biología de tiburones y rayas demersales en la región norte del Caribe Colombiano. *Acta Biológica Colombiana*, 13, 121-130.

Moreno, F., Acevedo, K., Grijalba-Bendeck, M., Polo-Silva, C. and Acero, A. (2009). Espectro trófico de la raya eléctrica Narcine bancroftii (Elasmobranchii, Narcinidae) en playa Salguero, Santa Marta, Caribe Colombiano. *Pan-American Journal of Aquatic Sciences*, 4, 413-422.

Moreno, F., Acevedo, K., Grijalba-Bendeck, M., Acero, A. and Paramo, J. (2010). Reproducción de la raya eléctrica Narcine bancroftii (Torpediniformes: Narcinidae) en Santa Marta, Caribe colombiano. *Latin American Journal of Aquatic Research,* 38, 27-35.

Munoz-Chapuli, R. (1984). Ethologie de la reproduction chez quelques requins de l'Atlantique nordest. *Cybium.*, 8, 1–14.

Musick, J. A. and Ellis, J. K (2005). Reproductive evolution of chondrichthyans. In Reproductive Biology and Phylogeny of Chondrichthyes: Sharks, Batoids, and Chimaeras, Vol 3 (Hamlett, W. C., ed), pp. 45-80. Playmouth, UK: Science Publishers, Inc.

Myers, R. A., Baum, J. K., Shepherd, T. D., Powers, S. P. and Peterson, C. H. (2007). Cascading effects of the loss of apex predatory sharks from a coastal ocean. *Science*, 315, 1846–1850.

Navia, A. F. and Mejía-Falla, P. A. (2006a). Aspectos biológicos, ecológicos y pesqueros de la fauna capturada incidentalmente en la pesca del camarón de aguas someras en el Pacífico colombiano. Documento Técnico No FS0206 Fundación SQUALUS. 42 pp.

Navia, A. F. and Mejía-Falla, P. A. (2006b). Dinámica poblacional y uso de hábitat y del tiburón aletiblanco (*Triaenodon obesus*), del tiburón ballena (*Rhincodon typus*) y de la mantaraya (*Manta birostris*), en el Parque Nacional Natural Gorgona, Pacifico colombiano. Documento Técnico No FS0306 Fundación SQUALUS. 27 pp.

Navia, A. F., Mejía-Falla, P. A., Caicedo J. A. and de Carvalho, M. R. (2006c). First record of Torpedo andersoni Bullis, 1962 in the western Caribbean region of Colombia (Elasmobranchii: Torpediniformes). *Caribbean Journal of Science*, 42, 231-233.

Navia, A. F., Giraldo, A. and Mejía-Falla, P. A. (2006d). Notas sobre la biología y dieta del toyo vieja (Mustelus lunulatus) en la zona central de pesca del Pacífico colombiano. *Investigaciones Marinas, Valparaíso*, 34, 217-222.

Navia, A. F., Mejía, P. A. and Giraldo, A. (2007). Feeding ecology of elasmobranchs fishes captured in the central fishing zone of the Colombian Eastern Tropical Pacific. *BMC Ecology*, 7, 1472-6785-7-8.

Navia, A.F., Mejía-Falla P.A. and Mejía-Ladino, L.M. (2008). Estado del conocimiento de tiburones y rayas En: INVEMAR. Informe del estado de los ambientes y los recursos marinos y costeros de Colombia. Año 2007. *Serie de publicaciones periódicas*, No 8. Santa Marta.

Navia, A. F., Mejía-Falla, P. A., Zapata, L. A., Bessudo, S., Soler G. and Rubio E. A. (2009). Estado del conocimiento de tiburones y rayas del Pacífico Colombiano. In Avances en el Conocimiento de Tiburones, Rayas and Quimeras de Colombia. (Puentes, V., Navia, A. F., Mejía-Falla, P. A., Caldas, J. P., Diazgranados, M. C. and Zapata, L. A., eds), pp. 197-245. Fundación SQUALUS, Ministerio de Ambiente Vivienda y Desarrollo Territorial, Instituto Colombiano Agropecuario, COLCIENCIAS, Conservación Internacional.

Navia, A. F., Cortés, E. and Mejía-Falla, P. A. (2010a). Topological analysis of the ecological importance of elasmobranch fishes: the Gulf of Tortugas food web, Colombian pacific ocean, as a case study. *Ecological Modelling,* 221, 2918-2926.

Navia, A. F., Mejía-Falla, P. A., López-García, J., L. Muñoz, A. and Ramírez-Luna, V. (2010b). Pesquería artesanal de la zona norte del Pacífico colombiano: aportando herramientas para su administración, Fase II. Documento técnico Fundación SQUALUS No. FS0110. 100 p.

Navia, A. F., Torres, A., Mejía-Falla, P. A. and Giraldo, A. (2011). Sexual, ontogenetic, temporal and spatial effects on the diet of Urotrygon rogersi (Elasmobranchii: Myliobatiformes). *Journal of Fish Biology,* 78, 1213–1224.

Navia, A.F., Cortés, E., Jordán, F., Cruz-Escalona, V.H. and Mejía-Falla, P.A. (2012). Changes to marine trophic networks caused by fishing. In Diversity of Ecosystems. (Mahamane, A., ed), pp. 417-452. Croatia: Intech.

Nelson, J. S. (2006). Fishes of the World. New York NY; John Wiley and Sons.

Perdomo, J. M. (2005). Rayas de agua dulce del genero Potamotrygon utilizadas como ornamentales de Colombia. Documento no publicado.

Payán, L. F., Mejía-Falla, P. A. and Navia, A. F. (2010a). Aportes al conocimiento de la historia de vida de la raya eléctrica *Narcine leoparda* en el Pacífico colombiano y validación de su estado de conservación a nivel local. Documento técnico Fundación SQUALUS No FS0310. 25 pp.

Payán, L. F., Mejía-Falla, P. A., Navia, A. F. and Lozano, R. A. (2010b). New records of Gorgona guitarfish Rhinobatos prahli in the Colombian Pacific Coast, *Marine Biodiversty Records*, 3, 1-3.

Payán, L. F., Navia, A. F., Rubio E. A. and Mejía-Falla, P. A. (2011). Biología de la raya guitarra Rhinobatos leucorhynchus (Günther, 1867) (Rajiformes: Rhinobatidae) en el Pacífico colombiano. *Latin American Journal of Aquatic Research,* 39, 286-296.

Polo-Silva, C. J. and Grijalba-Bendeck, M. (2008). Espectro trófico de la raya Guitarra Rhinobatos percellens (Walbaum, 1792) (Elasmobranchii: Rhinobatidae) en Santa Marta, Caribe, Colombia. *Memoria de la Fundación La Salle de Ciencias Naturales,* 169, 21-33.

Prada-Pedreros, S., González-Foreo, J. and Mondragón-Estupiñan, J. (2009). Capturas ícticas incidentales de la pesca ornamental en el período de aguas bajas en el área de influencia de Puerto Carreño, Orinoquia colombiana. *Universitas Scientiarum*, 14, 173-186.

Pratt, Jr. H. L. and Carrier, J. C. (2005). Elasmobranch courtship and mating behavior. In Reproductive Biology and Phylogeny of Chondrichthyes: Sharks, Batoids, and Chimaeras, Vol 3 (Hamlett, W. C., ed), pp. 129-169. Playmouth, UK: Science Publishers, Inc.

Punt, A. E., Pribac, F., Taylor, B. L. and Walker, T. I. (2005). Harvest strategy evaluation for school and gummy shark. *Journal of the Northwest Atlantic Fishery Science,* 35, 387–406.

Ramírez-Gil, H. and Ajiaco-Martinez, R. (2001). La pesca de especies de interés ornamental en el área de influencia de Inírida, Guainia. In La Pesca en la Baja Orinoquía Colombiana: Una Visión Integral. (Ramírez-Gil, H. and Ajiaco-Martinez, R., eds). pp. 139-154. Instituto Nacional de Pesca and Acuicultura, INPA.

Ramos-Socha, H.B. and Grijalba-Bendeck, M. (2011). bioecología de la raya de agua dulce Potamotrygon magdalenae (Duméril, 1865) (Myliobatiformes) en la Ciénaga de Sabayo, Guaimaral, Colombia. *Revista U.D.C.A Actualidad and Divulgación Científica*, 14, 109 - 118, 2011.

Rodríguez, M. A., Winemiller, K. O., Lewis, Jr. W. M. and Taphorn, B. D. C. (2007). The freshwater habitats, fishes, and fisheries of the Orinoco River basin. *Aquatic Ecosystem Health and Management*, 10, 140-152.

Rosa, R. S. (1985). A systematic revision of the South American freshwater stingrays (Chondrichthyes: Potamotrygonidae). Ph.D. Dissertation, College of William and Mary, Williamsburg: Virginia.

Rosa, R. S., Charvet-Almeida, P. and Quijada, C. C. D. (2010). Biology of the South American potamotrygonid stingrays. In Sharks and their Relatives II: Biodiversity, Adaptive Physiology, and Conservation (Carrier, J. C., Musick, J. A., and Heithaus, M. R., eds), pp 241-282. Boca Raton, FL: CRC Press, Taylor and Francis Group.

Robbins, W. D., Hisano, M., Connolly, S. and Choat, J. H. (2006). Ongoing collapse of coral reef shark populations. *Current Biology* 16, 2314-2319.

Salinas, Y. and Agudelo, E. (2000). Peces de Importancia Económica en la Cuenca Amazónica Colombiana. SINCHI, Instituto Amazónico de Investigaciones Científicas. Bogotá.

Shepherd, T. D. and Myers, R. A. (2005). Direct and indirect fishery effects on small coastal elasmobranchs in the northern Gulf of Mexico. *Ecology Letters,* 8, 1095–1104.

Springer, S. and Burgess, G. H. (1985). Two new dwarf dogsharks (Etmopterus, Squalidae), found off the Caribbean coast of Colombia. *Copeia,* 3, 584-591.

Stobutzki, I., Miller, M. and Brewer, D. (2001). Sustainability of fishery bycatch: a process for assessing highly diverse and numerous bycatch. *Environmental Conservation*, 28, 167–181.

Stevens, J. D. (2010). Epipelagic oceanic elasmobranchs. In Sharks and their relatives II: biodiversity, adaptive physiology, and conservation (Carrier, J. C., Musick, J. A., and Heithaus, M. R., eds). pp 3-37. Boca Raton, FL: CRC Press, Taylor and Francis Group.

Straube, N., Kriwet, J. and Schliewen, U. K. (2011). Cryptic diversity and species assignment of large lantern sharks of the Etmopterus spinax clade from the Southern Hemisphere (Squaliformes, Etmopteridae). *Zoologica Scripta,* 40, 61–75.

Téllez, L., Vargas, C. and Grijalba Bendeck, M. (2006). Algunos aspectos biológicos de Urotrygon Venezuelae Shultz, 1949 (Elasmobranchii, Rajuformes, Urolophidae), caprurada en playa Salguero, Santa Marta, Caribe de Colombia. Revista Científica U.D.C.A. *Actualidad y Divulgación Científica*, 9, 75-88.

Teshima, K. and Hara, M. (1983). Epithelial cells of the intestine of the freshwater stingray Potamotrygon magdalenae taken from the Magdalena River, Colombia, South America. *Bulletin of the Japanese Society of Scientific Fisheries,* 49, 1665-1668.

Teshima, K. and Takeshita K. (1992). Reproduction of the freshwater stingray, Potamotrygon magdalenae taken from the Magdalena river system in Colombia, South America. *Bulletin of the Seikai National Fisheries Research Institute,* 70, 11-27.

Thorson, T.B. (1972). The status of the bull shark Carcharinus leucas, in the Amazon River. *Copeia*, 1972, 601–604.

Villa-Navarro F. A. (1999). Estudio biológico pesquero de la represa de Prado para la determinación de especies promisorias en acuicultura. Presentado a Universidad del Tolima, Cortolima, Inpa, Gobernación del Tolima y Comité Departamental de Cafeteros del Tolima. Universidad del Tolima. Ibagué: Colombia.

Villa-Navarro, F. A., Zúñiga-Upegui, P. T., Castro-Roa, D., García-Melo, J. E., García-Melo, L. J. and Herrada-Yara, M. E. (2006). Peces del alto Magdalena, cuenca del río Magdalena, Colombia. *Biota Colombiana*, 7, 3-21.

Walker, T. I. (1998). Can shark resources be harvested sustainably? A question revisited with a review, of shark fisheries. *Marine and Freshwater Research,* 49, 553–572.

White, W.T. and Last, P.R. (2012). A review of the taxonomy of chondrichthyan fishes: a modern perspective. *Journal of Fish Biology*, 80, 901-917.

Wilson S. G., Polovina, J. J., Stewart, B. S. and Meekan, M. G. (2006). Movements of whale sharks (Rhincodon typus) tagged at Ningaloo Reef, Western Australia. *Marine Biology*, 148: 1157-1166.

Zapata, L. A., Acevedo, G., Angulo, W., Gómez, A., Acevedo J. and Gómez, G. (1998). Estudio biológico pesquero de las principales familias que conforman la pesca blanca en el Pacífico colombiano. *Boletín Científico INPA,* 6, 177-226.

Zeeberg, J., Corten, A. and de Graaf, E. (2006). Bycatch and release of pelagic megafauna in industrial trawler fisheries off Northwest Africa. *Fisheries Research,* 78, 186–195.

Electronic References

FAO. (2011). FAO fisheries and agriculture department. Available at: http://www.fao.org/fishery/statistics/global-capture-production/en (last accessed 12 July 2011).

Froese, R. and Pauly, D (2011). FishBase, world widc web electronic publication. Available at: http://www.fishbase.org (last accessed 12 July 2011).

IUCN, 2011. IUCN, International Union for Conservation of Nature. Available at: http://www.iucn.org/ (last accessed 12 July 2011).

In: Colombia
Editors: G. Lavigne and C. Cote
ISBN: 978-1-62417-848-1

Chapter 3

SOCIAL AND TECHNOLOGICAL ADVANCES FOR PEOPLE WITH DISABILITIES IN COLOMBIA

William R. Rodríguez[1,2]***, Daniel F. Jaramillo***[1]
and Gloria Del Rio[1]

[1]Stimulation, Leveling and Development Center - Cedesnid, Bogotá, Colombia
[2]University of El Rosario – Medicine School, Bogotá, Colombia

ABSTRACT

Having as a reference the work developed by the Center of Stimulation, Leveling, and Development CEDESNID for the habilitation/rehabilitation and social inclusion of the people with the condition of cognitive disability, this chapter has the purpose of doing a revision of the current status of cognitive disability in Colombia, to afterwards present the impact that can have the technological innovation in this field. For this it will be done a recount of the principal tendencies and lines of action at international level about the approach of the disability as a generality and of the cognitive/intellectual disability as a particular category. Afterwards, it will be made a revision of the most outstanding tendencies in Colombia. Finally, it will be presented the PreLingua tool as an example of the technological innovation of high social impact to be applied in the elimination of the involved barriers in the cognitive disability and as seed project for the innovation in Colombia.

1. INTRODUCTION

This chapter addresses some advances in social and technological environment for people with disabilities in Colombia, in especial, experiences in the *Stimulation, Leveling and Development Center CEDESNID*[*], a nonprofit institution that serves about 500 people in disability status or rights violation providing them protection and professional care. This work is focusing in two areas, on one hand, the social and educational inclusion of this kind of people, and on the other hand, technological changes and innovations related to voice therapy and early stimulation.

As professionals specialized in direct intervention in people with cognitive disabilities, the Cedesnid's team considers very important to know some aspects from patients like needs, daily live, accomplishments and futures goals in order to achieve better therapeutic interventions. In this way, the knowing, analysis and implementation of public policies governing cognitive disability in Colombia, as well as related policies in other developing countries, become in an important strategy which contributes to general well-being of the patients. Furthermore, one part that supports and complements the daily work of therapists is the technological part, thus, it is necessary to review different technical, scientific and social trends related to disabilities in order to benefit the patients, and at the same time, to understand the evolution of this phenomenon as a concept, as a social issue and as a phenomenon of knowledge. From this framework it is possible to deepen in the knowledge of the elements that link the needs of people with cognitive disabilities to academic and scientific advances and to social trends to satisfy these needs. Within this process it is necessary to understand proposals for action from the government and the scientific community in order to facilitate progress that integrates all of these different aspects and that results in the development and quality of life of people with this kind of disabilities. Also is important to mention that at this moment the Colombian Ministry of Health is in the process of reviewing public policy on disability for renewal, and that Colombia has ratified the UN Convention on the Rights of Persons with Disabilities, thus, is evident that now a days exist sufficient mechanisms to

[*] http://www.cedesnid.org.co/

satisfy the needs of this population in Colombia. At the same time, Colombia is experimenting positive changes in public policies on science, technology and innovation which promote initiatives and ideas with high social impact. About technological changes and innovations, Cedesnid is leading a project of research, development and innovation in free tools for voice therapy and multisensorial rooms for people with deep cognitive disabilities.

The project is founded by Colciencias, the Colombian entity that supports research and development initiatives in academic and business sectors. The aim of the project is to continue with the previous work of the computer-aided voice therapy tool called PreLingua. This tool is a set of free activities for providing interactive voice therapy to a population of individuals with voice disorders. The interactive tools are designed to train voice skills like: voice production, intensity, blow, vocal onset, phonation time, tone, and vocalic articulation for Spanish language. The development of these interactive tools along with the underlying speech technologies that support them requires the existence of speech processing, whose algorithms must be robust with respect to the sources of speech variability that are characteristic of this population of speakers. Currently the tool is used by more than 9200 therapist in 14 countries where the social impact is significant. In a previous study, PreLingua was applied in 27 patients of Cedesnid and the results show that the tool had certainly improved the voice skills of many of them according to the assessment of the therapists. However, the tool requires the improvement of its algorithms, the increase of its robustness, the development of an especial section in order to attend adult people and others tasks proposed by the therapist. All these improvements are considered in the mentioned project, also, the application of this technology in other areas related to special education like early stimulation and catch attention through multi-sensorial rooms. The above shows positive changes in Colombia in terms of research, development and social inclusion which contribute to real improvements in the quality of life of people with disabilities.

2. Disability Context

2.1. International Framework of Disability

While reviewing the current situation of disability, it is found there is great worldwide interest in improving the quality of life of the individuals with disabilities. The review finds views that highlight habilitation and rehabilitation processes, social inclusion and the recognition and exercise of the rights of such individuals, amongst others.

The WHO estimates that "500 million years of life are lost every year because of disabilities associated to health problems. The figure represents more than half of the years lost annually because of premature deaths" (Jiménez et al., 2002, p. 271). From this view point, there is a clear need to facilitate understanding disability, which may be useful when suggesting alternatives to approach and mitigate the social impact disability implies.

Studying and classifying disabilities allows knowing and preventing them in feasible cases or performing an early and adequate intervention in other cases. A good classification and description of the different disabilities permits experts to intervene and research accurately. It simultaneously leads social development and public policy creation towards higher prosperity scenarios for individuals with disabilities. From this perspective, knowledge about disabilities must outreach diverse academic, social and political groups in order to increase the development and social inclusion possibilities of this target population.

Among the different achievements made in this sense, one of the main references must be the International Classification of Functioning, Disability and Health (ICF), published by the World Health Organization and the Pan-American Health Organization (PAHO) in 2001. The ICF superseded the International Classification of Impairments, Disabilities and Handicaps ICIDH published in 1976, after the UN deemed it necessary to provide greater opportunities for individuals with disabilities (Jiménez et al. 2002).

In order to understand the appearance of the ICF, it must be considered that the ICIDH was based on three concepts: 1) *Impairments,* generally understood as body structure anomalies; 2) *Disabilities*, understood as the consequences of impairments; and 3) *Handicaps*, understood as the disadvantages suffered due to impairments and disabilities (Jiménez et al. 2002). This model was criticized because of its unidirectional approach focused on impairments, understanding disability as an individual phenomenon, and its scarce consideration of social and contextual variables. Facing such criticism and after a complex revision process (among other organizations, the Hispanic Language/Culture Network on Disabilities (RHHD) was created by several Hispanic countries for this process) the ICF publication took place (Jiménez et al. 2002).

The ICF was created as the first classification published in several languages (Jiménez et al. 2002) and it appears as a new science-based tool to focus on health studies (OMS/OPS/IMSERSO, 2001). Consequently, disability becomes an element directly related to health and awareness is raised on the significance of social and contextual aspects in the approach to disability. Disability changes from an individual issue and becomes a social issue defined as "a generic term including deficits, activity limitations and participation restrictions. It refers to the negative aspects of interaction between an individual with a health condition and their contextual factors (environmental and personal factors)." (OMS/OPS/IMSERSO, 2001, p. 231).

Pursuant to Jiménez et al. (2002), in 2002 the ICF had already been accepted by 191 countries, and one fundamental element was its intend for "substituting the negative perspective on impairments, disabilities and handicaps" by presenting a more positive and functional view of disabilities. This classification diverges from the ICIDH using a more positive model based on potential and functionality, getting closer to the International Classification of Diseases and Health-Related Problems ICD (Jiménez et al. 2002).

Subsequent to the ICF publication in 2007, the UN approved the Convention on Rights of Individuals with Disability and their Facultative Protocol, opening up new action and understanding scenarios related to disability. The convention focuses on acknowledging the human rights and fundamental freedoms of every individual with disability, reassuring their human condition as individuals with rights. Individuals with disability are defined in the convention as "individuals who have physical, mental, intellectual or sensorial long-term impairments, who interact with diverse barriers that may prevent their full and effective participation in society, under equal conditions with others" (United Nations, 2007, p. 4). Hence, we can see the Convention and the ICF share an approach where disability is related with the interaction of the individual with impairments and their social environment and the participation levels involved. Additionally, it is relevant to highlight that the convention suggests a vision where individuals with disability are equal to the rest of the society and it recommends a differential approach to women, boys and girls with disabilities (United Nations, 2007).

Parallel to these global contributions, actions tending to reassure the rights of the population with disability and facilitate development and social inclusion processes have progressed in different countries. Looking at Spain for example, Schalock and Verdugo's paradigm on quality of life applied to intellectual disability (further discussed below) and the proposal of the collective of individuals with disability stand out. The latter "has suggested the functional-diversity concept in order to demand their right to decision-making and end the marginalization they have been traditionally subject to". This position belongs to the independent-life assumptions and aims at superseding negative definitions (disability, handicap) to suggest the understanding of "disability" as one expression among many varieties that must be acknowledged currently in collective coexistence (Rodríguez et al., 2010). This is a current world trend focused on reducing names that may represent a negative condition or connotation in every inter-relation and performance at work, in family, society or personal life. Then, when addressing individuals with disability and other groups who may be excluded or whose rights may be violated, some social sectors are opting to strongly emphasize the use of new mechanisms to improve their quality of life.

On the other hand, in the south of Latin-America, based on Pierre Bourdieu's practical theory, Ferrante (2008) analyzed the position of individuals with disability in the "social space" in Argentina, using data from the 2002-2003 National Survey. The purpose was to find out to what extend the disability condition affects the position in "social space", considering the academic level reached and individual's activity level. From this sociological study where social, cultural and economic parameters were analyzed, he found that disability is related to positions in low social levels. This results in limitations to handle the physical space and own body. Hence, Ferrante (2008) questions the social dynamics involved in the phenomenon of disability, placing the discussion into an issue of social participation.

Proposals with similar elements to those shown above can be found in different regions. They fundamentally keep focusing on rights, quality of life and social participation, and simultaneously, there is a trend to find inclusive and generating languages. However, the approach to disability also has opposing views.

From a sociological perspective of disability, Barton (2008) gives more importance to human rights, social justice and non-discrimination; refuting the medical model because the diagnostic may label and stigmatize. The basic idea is a disqualifying diagnosis traditionally referring to terms such as "invalid", "crippled", "handicapped", "impaired" or "retard" (Barton, 2008). In the same sociological approach, Hahn (1986, quoted in Barton, 2008) states that disability is more related to failure in a social context unable to meet the needs of citizens with deficiencies than the disability of individuals.

Rieser and Mason (1990, quoted in Barton, 2008) question the medical model arguing that it has contributed to strengthen a wrong idea assuming that the deficiencies of individuals cause their disability. They also highlight that deficiencies may require medical assistance in many times, but this implies the need for the patient to be heard and their participation in the decisions affecting them.

In the same approach, Palacios (2008) states disability as a matter of human rights is an unquestionable reality in the current world. Palacios (2008) frames the understanding of disability within three fundamental perspectives: 1) a religious or "disregard" model derived from a belief of disability as a divine sign for society to disregard these individuals because they are not useful; 2) a rehabilitating model, here are scientific causes and the purpose is to rehabilitate and normalize individuals by hiding the difference; and 3) a social model that considers causes are neither religious nor scientific but social, and it suggests that individuals with disability can contribute to society similarly as those without disability. The principles are an independent life, non-discrimination, universal accessibility, environment normalization, civil dialogue, autonomy to decide about their own lives, elimination of any kind of barrier, and adequate balance of opportunities. This sociological perspective focuses more on social dynamics than the individual with disability; while other approaches, typically derived from health sciences, emphasize the individual's qualification and rehabilitation and the correction of something "bad". Consequently, diagnoses aid, medications, therapeutic interventions and even preventive or healing admissions are a fundamental part of this structure.

However, beyond the theoretical and methodological differences that may arise out of diverse approaches, the "rights" approach supported by the UN Convention and applied by different organizations working with disability is a reality today. Accordingly, speaking about rights, there is a reference to the participation of the individual with disability in decisions affecting them and thus their ability to choose. Although this model represents great progress in disability and social inclusion issues today, it is important to question cases where, in the social group's opinion, the individual lacks the ability to choose or at least such ability is reduced. Even though this statement might be rejected absolutely by some social groups working on disability, it is necessary to highlight that particular features, which are rarely considered when speaking of generic terms, are implicit in the so-called mental retardation, learning disability or intellectual disability as categories derived from or close to disability. Having this in mind some of the trends on this issue will be presented.

According to the American Association of Mental Retardation AAMR (2002), mental retardation is "a disability characterized by significant limitations both in intellectual functioning and in adaptive behavior, as evidenced by practical, social and conceptual skills. This disability is developed before the age of 18." (Luckasson et al. 2002, p8, quoted in Verdugo, 2004). From this view point, three main criteria are suggested to define the existence of mental retardation: intellectual functioning limitations, adaptive behavior limitations and start before 18. As stated by Verdugo (2003), the model suggested by the AAMR includes several dimensions:

- A dimension determined by the individual's intellectual abilities, meaning that different individuals have different intellectual abilities, especially when referring to intellectual disability.
- An adaptive-behavior dimension, related to the individual's possibility and ability to adapt to the development environment, depending on their functionality therein.
- A dimension related to participation, interactions and social roles.
- A dimension related to health including physical, mental health and the etiology of disability.
- A final dimension related to the context, considering the environment of development and culture.

The aforementioned dimensions are linked and supported by the aid the individual requires from the environment, aid that determines the individual's functioning. That is to say, individual's higher or lesser response reflected on the 5 dimensions will occur depending on the existing aids. Accordingly, it must be understood that pursuant to the definition suggested by the AAMR, intellectual disability is related to the interaction among personal, environmental factors, aid levels and expectations created around the individual and the situation (Verdugo, 2003). From this perspective, in order to be efficient, aid must be evaluated by identifying relevant areas first, and then, the relevant activities of each area and the level or intensity of each aid to finally achieve the individualized aid plan. The aforementioned procedure reinforces the idea of the need for a huge support network for this disability group that facilitates their life conditions.

From AAMR's conceptualization, it is remarkable that mental retardation is understood as a multidimensional disability related to intellectual functioning which requires generating different types of aids to improve the individual's functioning. Although this conceptualization has been accepted by several scientific and social communities, it is important to highlight that the AAMR changed its name to American Association on Intellectual and

Developmental Disabilities (AAIDD) some time ago, replacing the term "mental retardation" for "intellectual disability".

On the other hand, as previously mentioned, Spain applies Schalock and Verdugo's "quality of life" paradigm to the intervention of individuals with intellectual disability through an integral perspective (Verdugo et al. 2012). In this framework, it is understood that quality of life refers to a "desired state of personal well-being" (Verdugo et al. 2012, p. 13), which is multidimensional, has both universal features and features defined by culture, has subjective and objective components and is related to both personal and environmental characteristics (Verdugo et al. 2012). From this framework, it is suggested that the dimensions of quality of life are: "emotional well-being, inter-personal relationships, material well-being, personal development, physical well-being, self-determination, social inclusion and rights" (Verdugo et al. 2012, p. 13).

Something remarkable in this model in Spain is the investigation performed by the Excellence Research Group in Castilla y Leon GR 197 from 2009 to 2011, where six sub-projects were developed: 1) the creation of a quality-of-life integral scale; 2) early autism discovery and impact on the families' quality of life; 3) self-determination in mandatory high-school students; 4) the labor quality of life for workers risking social exclusion; 5) the quality of life in leisure for individuals with intellectual disability in Castilla y Leon; and 6) the quality of life and needs perceived by individuals with intellectual disability in the aging process in Castilla y Leon (Verdugo et al. 2012).

The quality-of-life integral scale was applied in several Latin-American countries such as Brazil, Argentina and Colombia; in addition to its applications in Spain. Very similar and coherent results in quality of life were found in the studied countries. Higher physical well-being scores were found in Argentina, interpersonal relationships in Brazil and self-determination rights in Colombia (Verdugo et al. 2012). However, the average places all these countries in a similar dimension of quality of life for individuals with disability, which suggests that the efforts being made have similar features.

Furthermore, the International Classification of Diseases (ICD) by the WHO, suggests mental retardation as "a disorder determined by the presence of an incomplete or stopped mental development, mainly characterized by damage to concrete functions in each development stage, those that contribute to intelligence, such as cognitive, language, motor functions and socialization" (quoted in Almenara et al. 1999, p. 383).

Another organization, the American Psychiatric Association, has developed a remarkable work. The association plans to define Intellectual

Developmental Disorder in DSM-5. In this sense, a change would be made from speaking about mental retardation –as used in the DSM-IV- to understanding it as an intellectual development problem with a deficit in the intellect and adaptive functioning arising during the development stage, not determined by IQ anymore, as it had been traditionally understood (American Psychiatric Association).

Upon gathering the up-to-date developments, it is noteworthy that approaches to disability move closer to multidimensional models in which the social component bears great importance. Similarly, it is remarkable that intellectual or cognitive disability is replacing mental retardation, as a category covering intellectual and adaptive limitations and framed within a social context playing a fundamental role in the approach. In this sense, the question about similarities and differences between disability as a global concept and cognitive or intellectual disability as a particular condition rise again. Considering these ideas about international scientific movements, a revision of the current Colombian scenario will be presented.

2.2. Colombian Framework of Disability

The Colombian Political Constitution from 1991 assigned the protection of all the marginalized groups to the State and defined specifically in article 47 the mandatory development of a social integration policy for individuals with disability. However, despite multiple efforts in this area, the national disability policy was not released until year 2004, based on a social approach model aiming at transcending the biomedical model (Consejo Nacional de Política Económica y Social – National Council of Economic and Social Policy - 2004). The policy is supported on environment integration, society and culture and it incorporates Social Risk Management (SRM) approaches and the Social Protection System (SPS). The purpose is to extend its outreach to individuals, families and communities, who are not necessarily living a situation of disability, but who are at risk of experimenting it. Consequently, disability and the risk of experimenting it can be understood as a series of conditions affecting the performance of an activity at some point in the living cycle (Consejo Nacional de Política Económica y Social, 2004). Evidently, the policy has a broad approach, using disability awareness as a generic constructor.

From this perspective, the Colombian State has focused on working to prevent new cases, mitigate disability effects and overcome limiting situations for individuals and the general population. Hence, emphasis has been made on guaranteeing "the entire population –not just few specific population groups- equal opportunities, risk management tools and interventions to overcome crisis situations" (Consejo Nacional de Política Económica y Social, 2004, p. 5). Additionally, it is noteworthy that the public policy on disability from 2004 recommended incorporating the ICF's approach, which evidences an early alignment with international proposals led by the WHO.

In line with these developments, Colombia has given great importance to adopting measures that guarantee the well-being and inclusion of individuals with disability. An example is set by progress in public policies of 3 main local cities such as Bogota D.C., Cartagena and Medellin. In 2007, the public policy on disability was adopted by the capital district. A complex and multi-cause definition of disability was taken here, and emphasis was made on inter-relational understanding over individual understanding. Additionally, it focused on the aim of integral development of the individuals with disability, their families and caregivers, being aware of the fact that the impact of disability does not affect the individual only, but their entire community. Finally, it is noteworthy that it has an approach of rights intended to facilitate social inclusion processes and improve the quality of life of individuals with disability (Alcaldía Mayor de Bogotá, 2007).

Subsequently, in 2010, the "Public Policy on Disability" document was released in the Cartagena District entitled "Keys for participation with social inclusion" 2010-2015. The suggested approach is based on acknowledging the human rights of individuals with disability and proposes to transcend strictly medical models to focus on abilities rather than on the absence or deficit (Alcaldía de Cartagena de Indias – Mayor's Office of Cartagena de Indias, 2010). Finally, in 2010, the public policy on disability was approved for Medellin in order to acknowledge its diverse population and guarantee the full exercise of human rights. In this sense, the policy was stated with 5 main components: 1) promotion and prevention; 2) balance of opportunities; 3) habilitation and rehabilitation; 4) communication and information; 5) Research and Training (Alcaldía de Medellín, 2010).

Simultaneously, local movements focusing on social inclusion for individuals with disability progressively appeared. The ratification of the Convention on Rights of Individuals with Disability in 2011 (supported by the REDDIS project) had a high national impact, becoming mandatory for all the public and private organizations of the Colombian territory. Hence, the approach to disability in Colombia has been in line with international movements. Such was the case that despite the lack of important elements as those mentioned so far, a redesign of the national public policy on disability is being currently developed with participation of the society with disability and its community.

In addition to socio-political aspects, it is important to know the Colombian situation of disability in terms of research. The systematization of research on Disability and rehabilitation strategies based on community from 2005 to 2010 (Cruz et al. 2011) provides a comprehensible scenario showing that although research about disability is increasing, greater efforts are still required. The study found that in Bogota (Colombian capital city) only 19 researches related to disability were conducted during the 5-year term and 11 were financed by the researchers themselves. As it can be seen, it is appealing that the country's capital city produces less than 3 researches on average per year; and even more appealing, more than 50% are financed by the researchers themselves (Cruz et al. 2011).

Despite the low amount of researches, it is important to mention that Cruz et al. (2001) highlight that most of the studies performed include the voices of individuals with disabilities as a fundamental element, which is in line with Gerber's ideas (2008 quoted in Cruz et al. 2011). Gerber states that their involvement is fundamental for them to be able to talk about their own lives. In this sense, a research tendency based on particular experiences of the population involved can be identified, which is consistent with a rights approach, seeking participation and social inclusion for individuals with disabilities.

In the research conducted by Hernández & Hernández (2005) with regard to disability in Colombia and its costs for the country, the bases are data estimated by the WHO, according to which 12% of the population would have some type of disability. Considering its 44 million inhabitants, this represents an estimated 5 million people with disability for Colombia and costs would be directly related to the disability seriousness (even though the National General Census for Disability of 2005 concluded that in Colombia the population with disability was the 6,4% (DANE, 2006), almost half of the figure proposed by the WHO, is still a very important proportion of the whole population). Consequently, higher seriousness would cause higher costs. The Hernández & Hernández`s research was conducted with data from the National Administrative Department of Statistics (DANE), which is the most reliable official source of statistics in Colombia, and different information parameters were used, categorized as follows: "1. The record for localization and characterization of individuals with disability (11); 2. The survey on quality of life 2.003 in relation to households of "individuals with permanent labor difficulties (IPLD)" (12); and 3. The largest possible series of the Continuous Household Survey from 1981 to 2003, regarding permanent labor disability (PLD) of the economically inactive population (13)" (Hernández & Hernández, 2005, p. 134).

The significance of this type of research for this paper and social development in Colombia lies in the fact that the strength the disability issue has started to acquire during the last year has become evident. It is a fundamental part of the country's public and academic agenda. Based on Hernández & Hernández's (2005) study, it was possible to find, for example, that "accidents, violence and conflicts are an important source of disability" and there is "economic exclusion of individuals with disability, and as a result their rights are neglected" (Hernández & Hernández, 2005, p.141). Another finding was that disability is mostly associated to poverty and old age, as data directly related to the individuals' quality of life.

Taking into account the guidelines of the public policy on disability and researches as the one quoted above, it is necessary to consider that, for example, the underprivileged population's poverty conditions are triggers of disability situations because they provide little possibilities to access health, sanitation, education, housing places and conditions, and quality of life in general. This increases participation limitations and the full enjoyment of their own rights; and simultaneously sets a scenario where the possibility to suffer new disabilities increases. In this sense, it is evident that extending the framework of disability understanding involves new challenges such as a reduction in poverty levels and generates greater labor expectations favoring individuals with disability both from the healthcare sector and the sociological work inherent to the topic and the relevant government institutions.

While integrating the disability with poverty issues, it is worth highlighting President Juan Manuel Santos`s administration, which has created the National Agency for Overcoming Extreme Poverty (ANSPE). This agency develops actions tending to eradicate extreme poverty and promote prosperity and social inclusion in the country.

In another demographic research performed by Cruz & Hernández (2008), interesting data about gender and types of disability in Bogota were found. At the end of the study, there was evidence of a greater disability effect on women than men. They also found higher frequency of disorders "related to body movement, present in 54.9% of individuals, followed by the nervous (48.9%), visual (43.8%), cardiorespiratory and immune (41.8%), digestive (23.9%), hearing (19.1%), voice and speech (17.3) system disorders of recorded individuals" (Cruz & Hernández, 2008, p. 26).

The ability to establish the percentage and effect of different types of disability represents some progress in the visualization of the problem, which guides adjustments, interests, definitions of support, studies, scientific advancements and policies related to disability in Colombia. The aforementioned progress is based on the statistical data found in Bogota, which being a capital city with a great population diversity represents a useful axis to guide future actions.

Accordingly, Moreno (2010) makes a collection of international and Colombian legislation related to disability and makes reference to each one. As far as Colombia is concerned, he finds a great deal of legislation especially from the 1991 Constitution, as well as, the jurisprudence addressed to the protection of the rights of individuals with disability. However, he mentions the scarce execution of standards, which places such individuals in a situation of neglected rights. Similarly, he alludes to the lack of clarity in the legislation in terms of the different attention levels required pursuant to the higher or lower commitment a disability might represent for each individual.

To quote another example which marks a demographic trend on the disability issue in Colombia, is the Gomez's (2007) study related to access to education. This study found that "the exclusion faced by individuals with disability is similar to that experimented by most Colombians; however, it is more evident due to their marginalization and segregation" (Quoted in Moreno, 2010, p. 18).

Upon introducing the studies above, instead of a discussion about the findings, it is pretended to provide an idea about the type of research being developed in Colombia in relation to disability and the social implications it may have. As evidenced before, the data obtained bear great national interest because they place the country in its current position and suggest work paths for the organizations involved in the topic. Nevertheless, after this understanding and action-trend revision related to disability as a general constructor, a revision of some remarkable elements around cognitive/intellectual disability as a particular category will be shown.

Pursuant to Law 1306 from 2009 of the Colombian State, "by virtue of which standards for the protection of individuals with mental disability are announced and the regime of legal representation for emancipated disabled is established", "a natural person has a mental disability when he suffers psychic or behavioral limitations not allowing him to understand the scope of his actions or leading to the undertaking of excessive or unnecessary risks in his asset management" (Bernal 2009, p. 1). Additionally, Law 1306 from 2009 states in article 8: "Individuals with mental disability shall have the rights of children and teenagers established by Title I of the Code of Childhood and Adolescence – Law 1098 from 2006…" (Bernal 2009, p. 5). In this sense, in order to understand social dynamics related to mental disability in Colombia – as a category close to intellectual disability or cognitive disability- it is important to highlight it is understood as a limitation implying dependency and/or risk. Consequently, individuals suffering this condition are protected by the same principles protecting children.

From this legal framework, the participation of all the individuals of a society in every sphere –including all the individuals with disability- stands out as a highly desirable situation. Jiménez (2011) states that "currently, individuals with disability are seen as individuals with rights, acknowledging and guaranteeing the balance of opportunities and participation in every area of community life (disability as a human rights issue), which implies fundamental roles and responsibilities for the environment to value and address human being diversity" (Jiménez, 2011, p. 1). In this sense, it is suggested that disability is diversity and inclusive education arises, adding great value to the social environment as responsible for the inclusion of an individual with disability. It is remarkable that the history of disability is a history of discrimination maintained by social policies, medical and rehabilitation models and education systems which must be transcended.

Mora et al. (2007) work in a close framework. In their study, they evaluated the quality of life of adult individuals with mental disability (called AIMD) and their caregivers, based on the modifications to life expectancy in the last 50 years. Among the most outstanding findings is the fact that individuals with mental disability in Cali are reaching older ages than before; as well as the fact that there is a lack of clarity in disability diagnosis, although most cases were early diagnosed. Similarly, a lack of guidance and clarity about possible aids AIMDs may access stand out. This leads to ideas of impossible change of their condition.

The most common service in healthcare findings was medication and there was a very low percentage of access or permanence in education. Basic performance was found in the labor aspect outside the household for men and domestic work for women, that is to say, activities depend a lot on the gender, in which women have less access to some kind of social contact related to their confinement inside the household. Additionally, it is noteworthy that incomes found –if existing- are lower than the minimum wage (Mora et al. 2007). Pursuant to these results, it can be seen that in important Colombian cities such as Cali, the quality of life of individuals with disability does not report the highest standards. However, the approaches made based on the quality-of-life concept to be applied to cognitive disability are highly significant.

In contrast to the previous research, there is a study by Arroyave y Freyle (2009), as a proposal to increase self-determination in teenagers with intellectual disability, by using a virtual scenario and a simulation program in order to obtain answers in "execution, decision-making and problem-solving" (Arroyave y Freyle, 2009, p. 58). Upon the relevant training under a controlled environment and applying a self-determination scale, a 30%·increment was found -as a result from the suggested training- in the possibility of expressing their emotions and feelings and making decisions in their everyday life as alternative functioning to adults' imposition. In addition to introducing an approach to reassure the individual with disability as a capable being, this research is a clear way to take advantage of IT and communications to support individuals with Cognitive/Intellectual Disability.

As seen in Colombia, there is a clear trend towards the inclusive approach of individuals with disability. On this regard, Bernal (2009) highlights the importance of strengthening an inclusive educational model by modifying the old disability concept of disability as a synonym of need for charity and pity, in order to facilitate positive understanding of productivity, inclusion and reaffirmation of rights.

Bernal (2009) mentions that different developments have been fundamental in this transition to offer individuals with disability a dignified life, support the development of their potential and their use of the aids provided by society.

Some of these developments are: 1) The 1991 Constitution (for the protection against abuse and the declaration of fundamental rights); 2) the General Law on Education from 1994 (with social and academic integration of all the children and teenagers); 3) Decree 2082 from 1996 with curricular adjustments and teacher specialization; 4) Law 361 from 1996 with mechanisms for the social integration of individuals with limitations; 5) Law 762 from 2002, which adopts measures to avoid discrimination against individuals with disability; 6) Decree 470 from 2007 of public policy on disability for the Capital District; 7) Decree 366 from 2009 with regulations of the pedagogical support service for individuals with disability and exceptional talented persons; 8) Law 1346 from 2009 approved by the convention on rights of individuals with disability adopted by the UN in 2006.

As we have seen, it is important to highlight that the Colombian tendency regarding the approach to intellectual/cognitive disability integrates a perspective that protects the individuals who experience it. Simultaneously, it tries to generate alternatives to improve their quality of life. Similarly, suggested viewpoints aim at emphasizing ability and diversity over lack and need for normalizing the situation, focusing a lot on the social tools that facilitate participation and full enjoyment of rights.

3. Technological Advances

Many changes are taking place in Colombia in technological contexts. For instance, the Royalties General System has recently change to consider now royalties for science, technology and innovation. The Fund for Science Technology and Innovation was created as a new fund to financing programs for regional projects of Science, Technology and Innovation allocating the 10% of the resources that Colombia receives for exploitation of non-renewable natural resources. Also, some institutions like the World Bank supports and funds Colombian entities like Colciencias in order to promote the social and technological advances.

3.1. Colciencias Grants

Job Placement Program for Doctors to Colombian Companies

Colciencias is the Administrative Department of Science, Technology and Innovation[1], the entity that supports the research and development initiatives in academic and business sectors in Colombia. With the support of the National Fund of Funding for Science, Technology and Innovation, Colciencias launched the 535 call: *Job Placement Program for Doctors to Colombian Companies* in 2011. The main idea of this initiative was integrate a PhD professionals into Colombian companies in order to promote and develop technological-based projects for three years. After the Colciencias review process, 23 projects (46%) were selected in that call. One of the selected project is related to the research, the development of technology and the innovation in tools for people whit disabilities, in special, computer-aided tools for voice therapy, early stimulation and catch attention. The project is already executing in the non profit institution Cedesnid and will benefit more than 400 people whit disabilities in that institution.

[1] www.colciencias.gov.co.

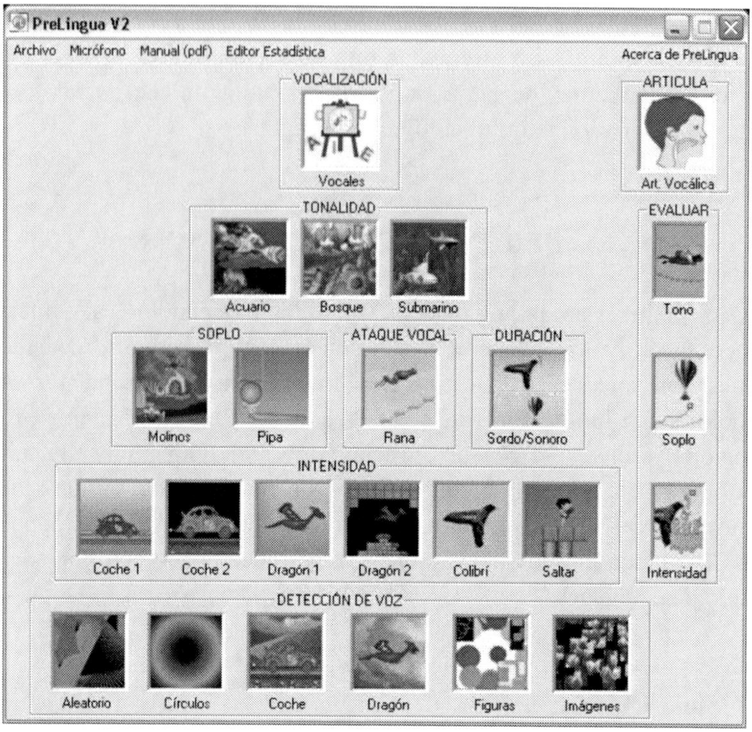

Figure 1. PreLingua - Tool for voice therapy.

3.2. The Seed Project

The project approved by Colciencias takes as seed initiative one project applied in children with voice disorders in spanish speaker-countries. That project was created under COMUNICA[2] initiative (Rodríguez et al., 2008), a set of free tools for voice and speech therapy designed for the *Voice Input Voice Output Laboratory*[3] in the Communication Technology Group from University of Zaragoza in Spain. In COMUNICA, all the computer-aided tools developed are free and include tools from the pre-linguistic stage until tools for speech therapy in the pragmatic level (Saz et al., 2008). One of these tools called PreLingua (Rodríguez and Lleida, 2008) was designed to train pre-linguistic skills and to control voice parameter in children with voice disorders, this tool can be seen in *Figure 1.*

[2] www.vocaliza.es
[3] http://vivolab.es/

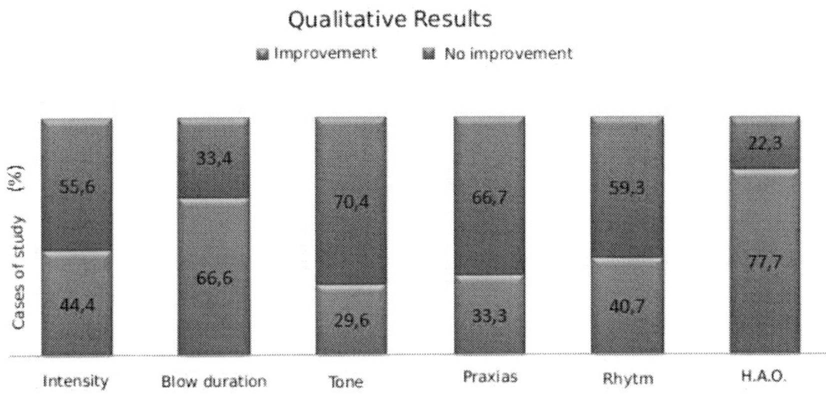

Figure 2. Qualitative results of study.

PreLingua was designed as a pyramid with five levels according to five rising degrees of difficulty in the activities. The base (level 1) corresponds to Voice Activity games, level 2 to the control of Intensity, level 3 to Blow, Vocal Onset and Phonation Time, level 4 corresponds to the Tone activities and, finally, level 5 to vocalic articulation activities with Vocalization and Articula (Rodríguez, 2011). The development of these kind of interactive tools requires joint technology like speech technologies, algorithms for speech processing (Rodríguez et al., 2008) (Saz et al., 2010) and a graphic engine[4], all in real time in order to give a visual feedback to the user of how to change his voice acoustical parameters.

In the context of disability is well known that working with altered child's voice is a difficult task by speech therapist, even more for those who work with impaired children. Therapists of institutions which offer aid to impaired people in special education schools, despite knowing the benefits of working hand in hand with technology, experience considerable needs and constraints in dealing with altered voice of this kind of population. On the one hand, by the few tools available that have a high acquisition cost, and on the other hand, due most of them are available in other languages than Spanish. Is here where PreLingua becomes in a useful tool due to its free distribution and its easy and understandable interface. In the PreLingua's research was necessary make an study in real patients in order to obtain measurable results and to test this technology. The study was conducted in special education centers in Spain and

[4] http://alleg.sourceforge.net/

Colombia for three months and included an objective assessment given by the system and a subjective assessment given by the centers' therapist.

Figure 2 shows the outcomes of the study where a number of the subjects who participated in the study using PreLingua had certainly improved the voice skills of many of them. Elements like blow, intensity and rhythm, which were improved after the therapy by 66.6%, 44.4% and 40.7% of the subjects, respectively, according to the therapists' assessment, were largely influence by the continuous work with PreLingua. Aspects like Tone and Tongue Praxis has less relevant results due to the high concentration level required, and finally, an additional column (Habilidades Adicionales Observadas in spanish) with topics highlighted by the therapist as additional skills, this include a high motivational attitude, increment of general attention and healthy competition between the users.

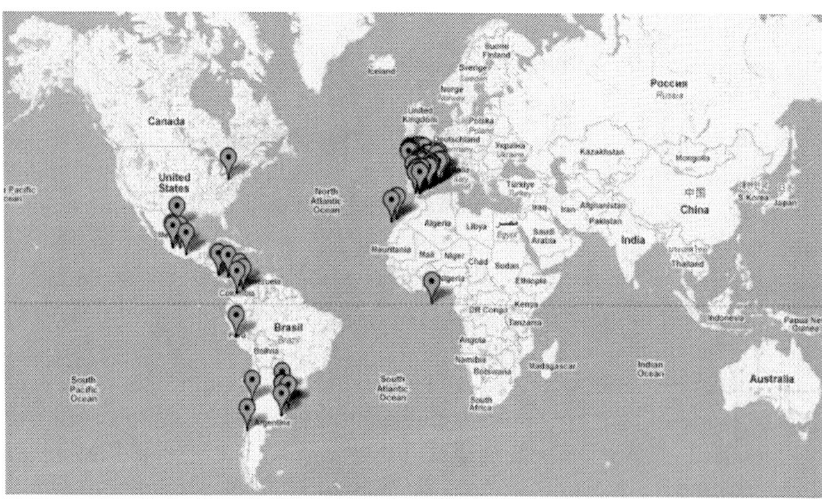

Figure 3. Global users of PreLingua.

3.3. Colciencias Project

A Project with High Social Impact

Nowadays, PreLingua is used by more than 9200 therapists in 14 countries as is shown in *Figure 3* where the social impact is significant. The Colciencias project will continue the work already begun with PreLingua, moreover, by considering the results of its application in the study mentioned above, it is clear the high potential of benefit for people with disability in Latin-America (spanish speaker countries). Many suggestions and comments received from therapists of different countries will be considered in the future version of the tool which is developing in Colombia. New features like multi-platform installers, improvement of its algorithms, the increase of its robustness, the development of an especial section in order to attend adult people and others tasks will be included in the next version of the tool. All this in order to strengthen research and innovation in therapeutic techniques to improve the quality of life of people with disabilities.

The currently project founded by Colciencias represents for Cedesnid and for people with disabilities in Colombia a tangible advances in technological and social contexts. It is the beginning of many and positives changes what will surely bring other projects and initiatives founded by government entities or NGO. One of the keys to success this kind of project consist in the collaborative work between different actors like engineers, therapist, parents and people with disabilities themselves in order to ensure real technological solutions for people with real needs.

CONCLUSION

Disability has generated diverse concepts and positions throughout time, both individually and collectively. It has traveled through time from the horror and disdain of early cultures, going through the disregard and abuses from some years ago to reach this crucial time of dignifying individuals who require their rights to be observed and their existence to be acknowledged due to different reasons.

The international movement generated in the last decades in favor of individuals with disability is increasingly stronger and more human because social trends are progressively and conclusively leading to the acknowledgement and respect of the differences of minority groups, among which disability stands out due to the human potential wasted and trashed throughout centuries.

In this line of thought, Colombia enters as a country determined to demand the rights of individuals with disability and generate strategies which integrate the entire population instead of separating it. However, for this purpose to become true in every corner of Colombia, there is still a lot of work to do both on the conceptual field to facilitate understanding applicable to particular situations and the technological field to develop tools to eliminate the barriers that may limit participation.

References

Alcaldía de Cartagena de Indias. (2010). *Política Pública en discapacidad en el distrito de Cartagena - Claves para la participación con inclusión Social 2010–2015*. Retrieved Sep 2, 2012, from: http://www.cartagena.gov.co/Concejo/Documentos/CARTILLADISCAPACIDAD.pdf

Alcaldía de Medellín. (2010). *La Política Pública y el plan Municipal de Discapacidad 2010 – 1018 del Municipio de Medellín: "Construyendo juntos una ciudad para todos: Hacia una Medellín accesible e incluyente"*. Retrieved Sep 2, 2012, from: http://www.medellin.gov.co/irj/go/km/docs/wpccontent/Sites/Subportal%20del%20Ciudadano/Bienestar%20Social/Secciones/Programas%20y%20Proyectos/Documentos/2011/documentos%20interes/informe%20politica%20publica%202011.pdf

Alcaldía Mayor de Bogotá. (2007). *Decreto No. 470 del 12 de octubre de 2007*. Distrital Record 3855, Oct 12, 2007.

Almenara, J., García, R., Novalbos, R., Merello, B., Abellan, M. & García, C. (1999). *Evaluación Médica y Psicosocial de una Población Adulta con Discapacidad Intelectual*. Revista Española de Salud Pública. 73: 383 – 392.

American Psychiatric Association (n.d). *DMS-5 development*. Retrieved Oct 2, 2012, from: http://www.dsm5.org/Pages/Default.aspx

Arroyave, M. & Freyle, M., (2009). *La autodeterminación en adolescentes con discapacidad intelectual*. Innovar, Especial en Educación. Dic. 2009, pp. 53 – 64.

Barton, L. (2008). *Sociología y Discapacidad: algunos temas nuevos*. En: Barton, L. (comp.), *Discapacidad y Sociedad*. Madrid, Morata Ed.

Bernal, C. (2009). Congreso de la República de Colombia. *Ley 1306 de 2009. Legislación que favorece la educación inclusiva*. Retrieved Sep 25, 2012, from:
http://www.corporacionsindromededown.org/userfiles/Legislacion.pdf

Consejo Nacional de Política Económica y Social (2004). *Documento CONPES Social 80. Política Pública Nacional de discapacidad*. Bogotá: Departamento Nacional de Planeación.

Cruz, I. & Hernández, J. (2008). Magnitud de la Discapacidad en Colombia: una aproximación a sus determinantes. *Revista Ciencias de la Salud*. 6 (3): 23 – 35.

Cruz, I., Duarte, C., Fernández, A. & García, S. (2011). *La Sistematización de Investigaciones en Discapacidad y en la Estrategia de rehabilitación basada en Comunidad periodo 2005 – 2010.* Alcaldía Mayor de Bogotá D.C.

Ferrante, C. (2008). Cuerpo, Discapacidad y posición Social: Una aproximación indicativa al habitus de la Discapacidad en Argentina. Intersticios: *Revista Sociológica de Pensamiento Crítico.* Vol. 2 (1) pp. 173 – 185.

Hernández, J. & Hernández, I. (2005). *Una aproximación a los costos indirectos de la Discapacidad en Colombia.* Revista de Salud Pública. 7 (2): 130 – 144.

Jiménez, D. (2011). *Concepto Social de la Discapacidad.* Secretaría de Educación de la Alcaldía Mayor de Bogotá. Retrieved Sep 29, 2012, from: http://www.corporacionsindromededown.org/userfiles/Concepto%20social%20de%20la%20discapacidad%20Patrica%20Jimenez.pdf

Jiménez, M., González P. & Martín J. (2002). *La Clasificación Internacional del Funcionamiento y de la Salud.* Revista Especializada de Salud Pública. No. 4, pp. 271 – 279.

Mora, A., Córdoba, L., Bedoya, A. & Verdugo, M. (2007). *Características de la calidad de vida en familias con un adulto con discapacidad intelectual (DI)/RM en la ciudad de Cali, Colombia.* Revista Diversitas – Perspectivas en Psicología. Vol. 3, No. 1, pp. 37 – 54.

Moreno, M. (2010). *Infancia, Políticas y Discapacidad.* Bogotá: Miller impresores.

United Nations. (2007). *Convención sobre los Derechos de las Personas con Discapacidad.* Retrieved Sep 24, 2012, from: http://www.un.org/spanish/disabilities/default.asp?id=497

OMS/OPS/IMSERSO (2001). *Clasificación Internacional del Funcionamiento, de la Discapacidad y de la Salud: CIF.* Grafo S.A. Ed.

Palacios, A. (2008). *El Modelo Social de Discapacidad: Origenes, caracterización y plasmación en la convención internacional sobre los derechos de las personas con discapacidad.* Madrid, CINCA, Ed.

Rodríguez, S. & Ferreira, M. (2010). *Diversidad Funcional: Sobre lo normal y lo patológico en torno a la condición social de la discapacidad.* Cuadernos de Relaciones Laborales. 28 (1) pp. 151 – 172.

Verdugo, M. (2003). *Aportaciones de la definición de Retraso Mental (AAMR, 2002) a la corriente inclusiva de las personas con discapacidad.* Retrieved Sep 2, 2012, from: http://ruta000.com.ar/inspeccion/wp-content/uploads/2011/10/Verdugo-Alonso-Discapacidad-Intelectual.pdf

Verdugo, M. (2004). *Cambios Conceptuales en la Discapacidad. II Congreso Internacional de Discapacidad Intelectual: Enfoques y Realidad: Un Desafío*. Salamanca: Publicaciones del INICO.

Verdugo, M., Canal, R., Jenaro, C., Badía, M. & Aguado A. (2012). *Aplicación del paradigma de calidad de vida a la intervención con personas con discapacidad desde una perspectiva integral*. Salamanca: Publicaciones del INICO.

DANE (2006). *Censo General 2005. Discapacidad. Personas con limitaciones permanentes,* Departamento Administrativo Nacional de Estadística, Retrieved Oct 9, 2012, from: http://www.dane.gov.co/ files/censo2005/discapacidad.pdf.

Rodríguez, W. and Lleida, E. (2008). *PreLingua una herramienta para el desarrollo del pre-lenguaje*. In V Jornadas en Tecnologías del Habla, Bilbao, Spain.

Rodríguez, W., Vaquero, C., Saz, O., and Lleida, E. (2008). Speech technology applied to children with speech disorders. In: *Proceedings of the 4th Kuala Lumpur International Conference on Biomedical Engineering*, Kuala Lumpur, Malaysia.

Rodríguez, W., (2011). *Tecnologías del Habla en la Educaciónn de la Voz Infantil Alterada: Cómo las TIC pueden ayudar a educar la voz alterada en niños con o sin discapacidad*, Madrid, Spain, Editorial Académica Española.

Rodríguez, W., Saz, O., Lleida, E., Vaquero, C., and Escartín, A. (2008). *Comunica - tools for speech and language therapy*. In: Proceedings of the 2008 Workshop on Children, Computer and Interaction, Chania, Greece.

Saz, O., Lleida, E., Rodríguez, V., Rodríguez, W., and Vaquero, C. (2010). *The use of Synthetic Speech in Language Learning Tools: Review and case study*. In: Mullennix, J. and Stern, S. (ed) *Computer Synthesize Speech Technologies,* p. 188-20. USA: IGI Global Publishing.

Saz, O., Rodríguez, W., Lleida, E., Vaquero, C., and Escartín, A. (2008). *Comunica - plataforma para el desarrollo, distribución y evaluación de herramientas logopédicas asistidas por ordenador*. In: Proceedings of V Jornadas en Tecnologías de Habla, Bilbao, Spain.

In: Colombia
Editors: G. Lavigne and C. Cote

ISBN: 978-1-62417-848-1

Chapter 4

CLIMATE AND NATURAL HAZARDS ON THE COLOMBIAN CARIBBEAN REGION

Juan Carlos Ortiz Royero*
Area of Ocean and Atmosphere - Applied Physics Group – Physics Department. (Área de Océano y Atmosfera - Grupo de Física Aplicada- Departamento de Física). Universidad del Norte. Barranquilla, Colombia

ABSTRACT

This chapter describes the climate of the northern Colombia called the Colombian Caribbean region. The climate in this area of Colombia is driven by the movement of the inter-tropical convergence zone (ITCZ), which produces two seasons: a dry (low rainfall) season from November to May and a rainy season (high rainfall) from April to October, with a season of transition called 'veranillo de San Juan' between June and July. Also, the vulnerability to natural hazards is described; with emphasis in phenomena such as hurricanes and storms, tornadoes and cold fronts. Although the Colombian Caribbean is not known as a natural disaster area, hurricanes (e.g., Joan in 1988 and Lenny in 1999), tornadoes, windstorms and cold fronts, can affect the climate of this Caribbean area, causing damage to infrastructure and loss of life in some cases.

* E-mail: jortiz@uninorte.edu.co.

INTRODUCTION

The surface area of the Colombian Caribbean is close to 590,000 km^2, the Caribbean Sea is bounded by South America to the south, Central America to the west, the Lesser Antilles to the east, and the Greater Antilles (Cuba, Haiti, Dominican Republic, and Puerto Rico) to the north. It is connected to the Gulf of Mexico by the Yucatan channel to the northwest and to the tropical Atlantic Ocean to the east. The Colombian Caribbean coast is approximately 1,600 km long, and the main coastal cities in this important region are Riohacha, Santa Marta, Barranquilla, and Cartagena.

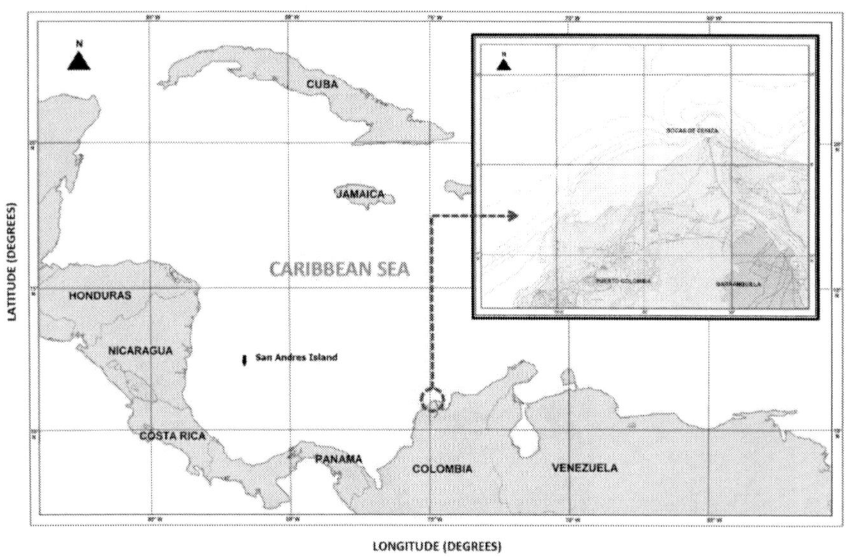

Figure 1. Colombian Caribbean Sea and San Andrés Island (Ortiz et al 2012).

These sea and river port cities are of great economic importance for Colombia's economy. San Andrés is located in the Caribbean Sea, approximately 192 km to the east of Nicaragua, northeast of Costa Rica, and 775 km to the northeast of the Colombian coast. San Andrés Island is the largest of the islands in the Archipelago of San Andrés, Providencia, and Santa Catalina and it has a total area of 26 km^2. It has belonged to Colombia since 1803. The northeastern part of the island is surrounded by a small coral reef and several cays that are home to a variety of marine flora and fauna, which attract crowds of tourists every year (approximately 350,000). San Andrés has an approximate population of 75,000 inhabitants; Figure 1 shows the Colombian Caribbean and the location of the island of San Andrés. (Meisel, 2003).

The former regions amount to 11.6% of Colombia's entire national territory of 1,141,748 km^2. Barranquilla is the capital of the Atlántico Department and is located close to the Magdalena River delta in the Caribbean Sea. The Magdalena River is the longest in the country, with a total length of 1,613 km. The river has an average flow volume of 7,000 m^3/s and an approximate yearly discharge of 200 million tons of sediment (Alvarado, 2009).

Barranquilla, which is located at 74° 48' W and 10° 58' N, is one of the most important and active ports, and it is the main commercial, industrial, cultural and academic center of the Colombian Caribbean region. Barranquilla has a population of 1,148,506, thus making it the fourth most populated city in the country after Bogota, Medellin and Cali. The Figure 2, show the port entrance of Barranquilla.

Figure 2. Barranquilla port entrance (Image taken from Google Earth® 2012).

CLIMATE

According to Invemar (2010), the climate of Colombia's Caribbean coast is determined by its tropical location, which exposes it to direct solar radiation. Accordingly, there are no defined climactic seasons, in contrast to those latitudes further from the Equator. In this region, there are two seasonal periods determined by precipitation and the patterns of the Aloysius Winds of the East.

As it is shown on Figure 3, the movements of the Inter-tropical Convergence Zone (ITCZ) explain these seasons, there is a dry period with strong winds (December to April), a transitional period (May to July) and a rainy period (August to November). These three climactic periods may change both in duration and in intensity due to the influence of the American Monsoon system, low-level atmospheric wind currents and the 'El Niño' and 'La Niña' (positive phase of 'El Niño') events (Andrade, 2001).

The main dry season in the Colombian Caribbean occurs from December to April and the wet season the rest of the year, interrupted by a relative minimum of precipitation in July and August known as 'veranillo de San Juan'. In the dry season, and the 'veranillo de San Juan', the winds are associated with the San Andrés jet and the northeast trade winds.

When the winds of San Andrés jet are stronger the dry season occurs throughout the Caribbean coast, which coincides with intense NE trade winds and the same goes for the 'veranillo de San Juan'. Instead, the wet season coincides with the period of greatest intensity of the Chocó jet winds and less intensity of the San Andrés jet winds. Throughout the entire Caribbean coast the wettest month of the year is October, and the driest are February and March, however, there is a strong spatial variation of this parameter (Bernal et al., 2006).

From December to March, the ITCZ is located over South America and the Caribbean high pressure centers are strong and the northeast trade winds are maximum, which is consistent with the main dry season in the Caribbean region. From July to September the location of the ITCZ shifts to the center of the Caribbean, the trade winds weaken and southwesterly winds are dominant, these winds are weaker but can bring strong storms to the Caribbean region (Andrade and Barton, 2001). These annual cycles coincide with peaks in the San Andrés jet and Chocó, respectively. However, the second peak of San Andrés jet, which occurs in July-August associated with the Indian summer, is not directly related to the ITCZ but a temporary intensification of the high pressure system in the North Atlantic.

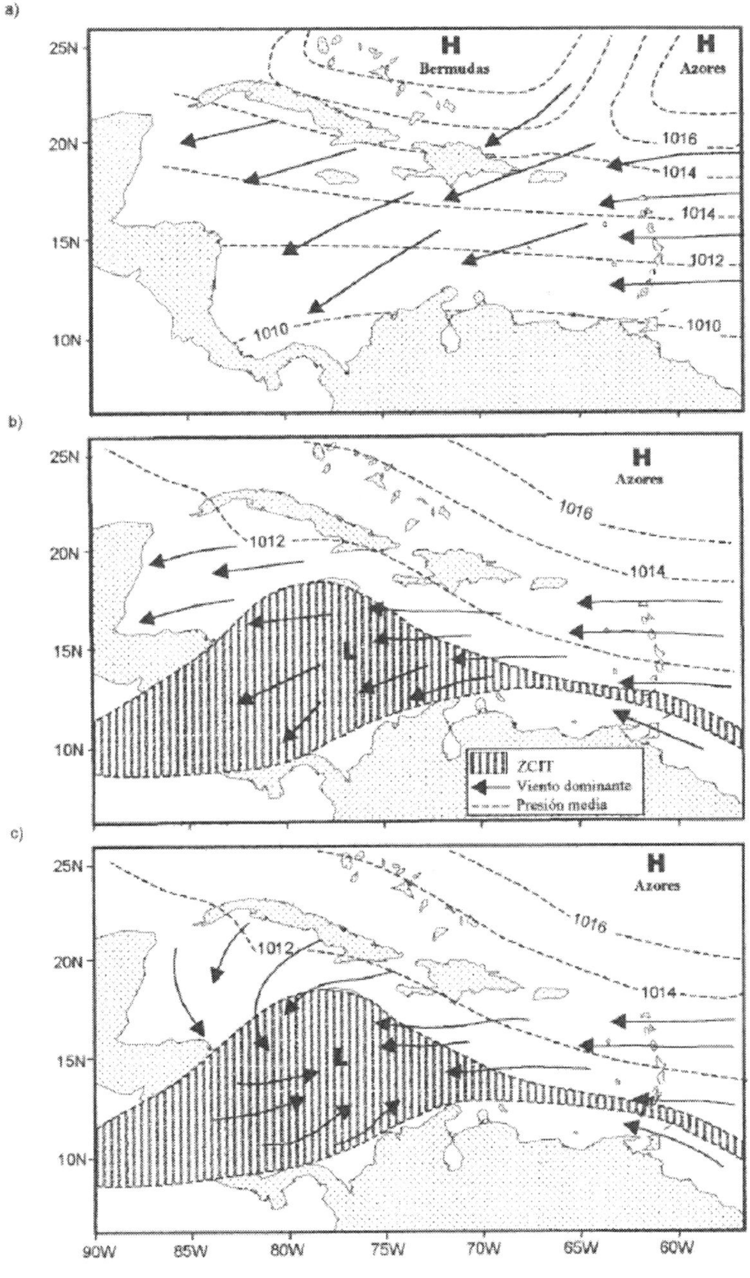

Figure 3. Winds vectors and contours of atmospheric pressure in the Colombian Caribbean Sea during dry (a), Intermediate (b) and wet seasons (c), (Andrade, 2001).

The mean temperature in the area, obtained by a multiyear average on data from airports's meteorological stations is around 28.5°C (Figure 4). The relative humidity of the study area is about 73.6% (Figure 5).

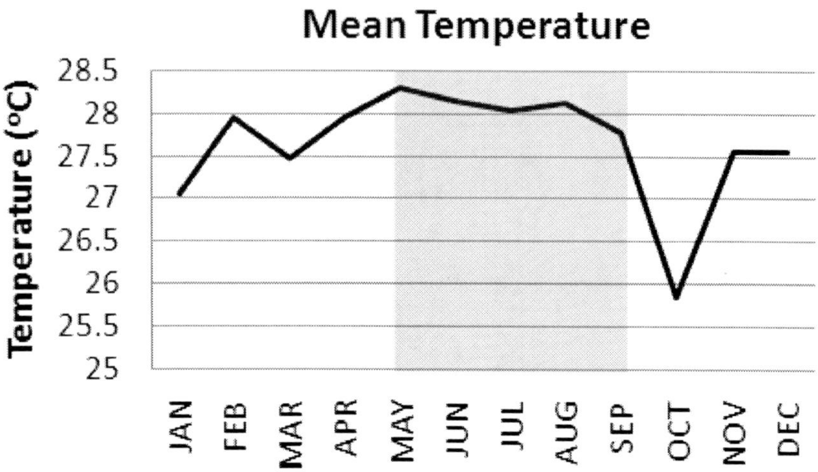

Figure 4. Mean temperature of the Colombian Caribbean zone (Ortiz and Rosales, 2012).

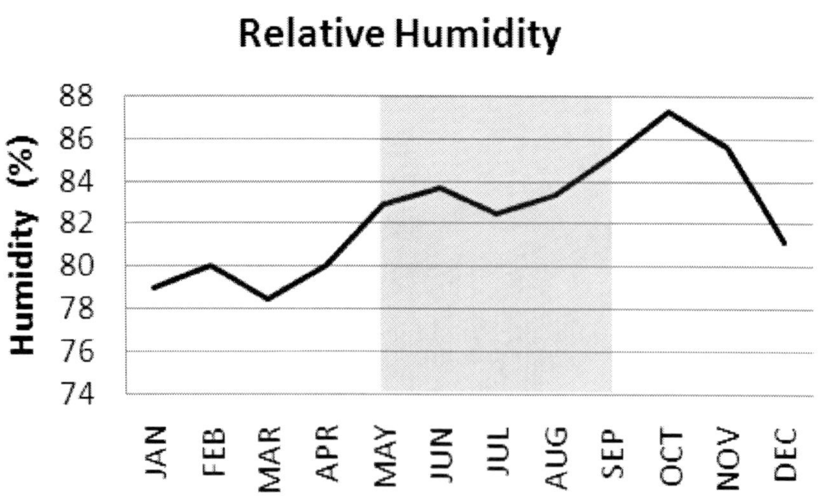

Figure 5. Mean humidity of the Colombian Caribbean zone (Ortiz and Rosales, 2012).

The higher values of temperature are observed in the period May-September, with an average of 29.5°C and going as low as 27°C. Sometimes the temperature during September reaches the 35 – 40 °C range. In the period December to April, monthly averaged temperatures between 27 and 28.0°C are observed. For the wet season the trend of this parameter is to decrease, due to noticeable increases in cloudiness, maintaining a monthly average temperature between 28°C and 28.5 ° C. The highest values of this parameter are observed in the transition period and the minimum at the beginning and end of the year.

The driest time of the year is the period between mid-December to mid-April. The rainy seasons are mainly from mid-April to mid-December, with a significant in precipitation during the months of June and July, marking the second dry season, which is wetter than the first.

Overall the area enjoys a climate with significant water stress (dry), influenced by the winds already mentioned. In general, the study area is characterized by temperatures above the 27° C and potential evaporation greater than precipitation, being these effect larger to the north of the study area. The multiyear average precipitation in the south is about 371 mms, while at the north goes down to 311 mms (Figure 6).

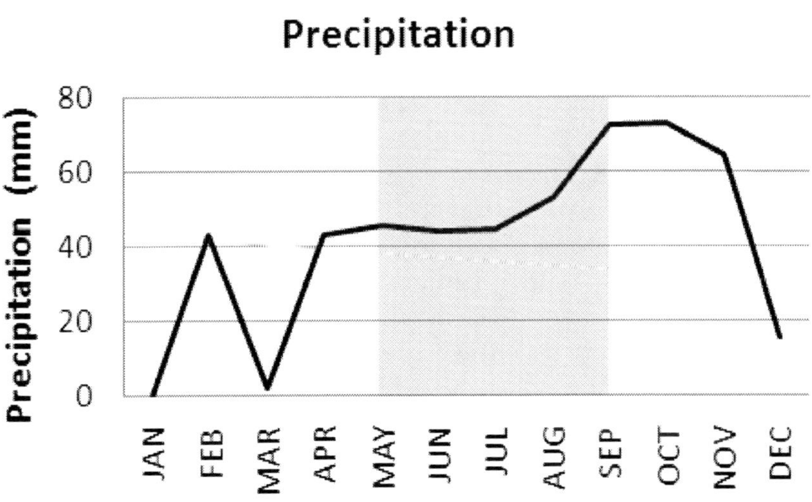

Figure 6. Mean precipitation of the Colombian Caribbean zone (Ortiz and Rosales, 2012).

The weather stations of the airports are consistent in their annual seasonality; it is typical to observe peaks in the wind speed in the dry season from December to February (Figure 7), a time of transition with a bit of wind relaxation named 'veranillo de San Juan' occurs in the middle of the year between July and August and the maximum rainfall and very light winds are observed from September to November.

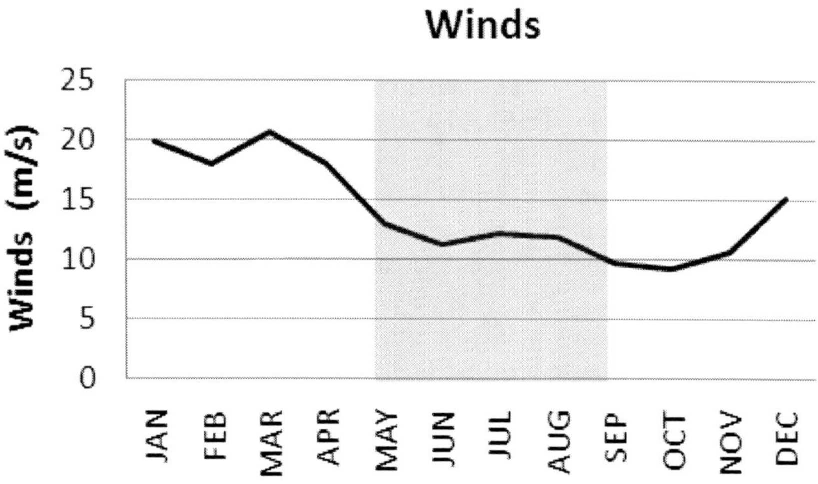

Figure 7. Mean winds of the Colombian Caribbean zone (Ortiz and Rosales, 2012).

The hurricane season for the Caribbean area starts at the beginning of June and runs through November, being statistical data from August and October the most intense.

The probability that these systems affect Colombian territory is very low compared to other areas of the Caribbean. Yet significant events have affected the Colombian Caribbean, so that the departments as Cesar, Córdoba, Magdalena, Atlántico, Bolivar, Guajira, San Andres and Providencia, may be affected, being the lattest two the most vulnerable. Hurricane Joan in 1988 is remembered as the most important event.

Natural Hazards

Although the Colombian Caribbean coast is not registered as a potentially vulnerable zone in international databases about of natural hazards, recent works show that there is an important impact of different phenomena, e.g., storms and hurricanes, cold fronts, tornadoes, an sea level rise, due to global changes.

Hurricanes

Irene in 1971, Joan in 1988, and Bret in 1993, are some examples of storms that crossed the Colombian coast in the past (Figure 8).

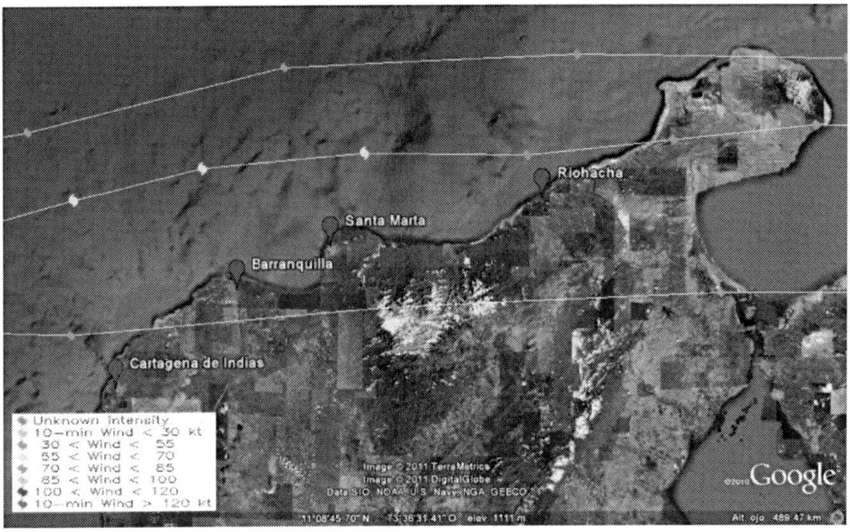

Figure 8. Hurricanes close to the Colombian Caribbean coast (Ortiz, 2012).

The increase in the number and intensity of the storms that passed through the Caribbean Sea in the last decade shows the need to begin to evaluate the effects of these phenomena in the coastal zone to design warning programs (Ortiz, 2007). The waves and the storm surge represent the most important oceanic effects due to an event of this nature. The former work does a review and analysis of the storms that crossed the Colombian Caribbean since 1900, their characteristics and the coastal threat that they represent. It is established that the last 17 years have been very More intense in storms and it is observed that the storms are becoming a real threat for the Colombian Caribbean coast; The increase in the number of events around 2005 is in agreement with the trend of increase in global temperature Experts in global warming agree in that the intensity or the number of storms in the Atlantic can increase due to the effect of the increase in sea surface temperature of the ocean.

It has been established that that there have been significant storm events in the region, with some of them hitting land. The analysis made possible to identify ten significant events that have occurred since 1900 and to determine the vulnerability of the Colombian territory to storms and hurricanes, Tables 1 and 2 show the most important hurricanes in Colombia and their distance to main cities of the Colombian Caribbean zone (Ortiz, 2012).

Table 1. Storms and hurricanes near the Colombian Caribbean coast

Name	Category	Date	Closest distance to coast (km)	Maximum wind (km/h)
Storm No. 10	Tropical Storm	1909-10-06	RIO (350.6)	55.6
			STA (205.5)	55.6
			BAQ (75)	55.6
			CTG (64)	74.1
Storm No. 4	Tropical Storm	1911-09-07	RIO (130.6)	101.9
			STA (208.3)	101.9
			BAQ (270)	101.9
			CTG (375)	101.9
IRENE	Tropical Storm	1971-09-17	RIO (108)	64.8
			STA (133)	64.8
			BAQ (160)	64.8
			CTG (228)	64.8
JOAN	Tropical Storm	1988-10-18	RIO (26)	111.1
	H1		STA (64)	120.3
	H1		BAQ (79)	129.6
	H1		CTG (116)	138.9
BRET	Tropical Storm	1993-08-09	RIO (120)	74.1
			STA (60)	74.1
			BAQ (46)	74.1
			CTG (107)	64.8
CESAR	Tropical Storm	1996-07-26	RIO (96)	74.1
			STA (115)	83.3
			BAQ (174)	83.3
			CTG (150)	92.6

Furthermore, it was possible to establish four areas of vulnerability as follows: area 1 (greatest vulnerability), the Archipelago of San Andrés; area 2 (medium vulnerability), the coastal zone of the Department of Guajira; area 3 (medium-low vulnerability), the coast of the Departments of Atlántico and Bolívar; and area 4 (low vulnerability), the coast of the Departments of Córdoba and Antioquia (Figure 9).

Table 2. Storms and hurricanes close to the Archipelago of San Andrés

Name	Category	Date	Closest distance to coast (km)	Maximum wind (km/h)
Storm No. 4	H1	1911-09-09	89	138.9
HATTIE	Tropical Storm	1961-09-17	60.1	101.8
ALMA	Tropical Depression	1970-05-18	131.1	46.3
JOAN	H3	1988-19-21	73.8	203.7
CESAR	Tropical Storm	1996-07-28	62.7	120.4
KATRINA	Tropical Depression	199-10-29	60.5	55.6
BETA	Tropical Storm	2005-10-28	70	92.6

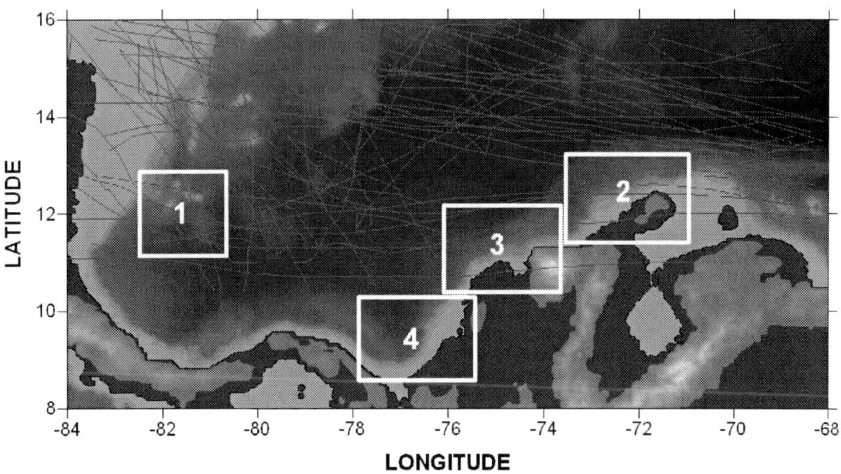

Figure 9. Vulnerability of Colombian Caribbean Sea associate to storms and hurricanes (Ortiz, 2012).

Two hurricanes are in record in the last 25 years in the Colombian Caribbean coast: Hurricane Lenny, and Hurricane Joan. On October 18 at 18:00 hrs (local time) Joan became a Category 1 hurricane some 50 km from the Colombian coast, with maximum wind speed of 120 km / h and 990 mb pressure. Just one day later, Joan had reached Category 2 with sustained winds of 167 km / h, and 976 mb pressure at 100 km from the Colombian coast, arriving in the area known as the Gulf of Darien (Figure 10).

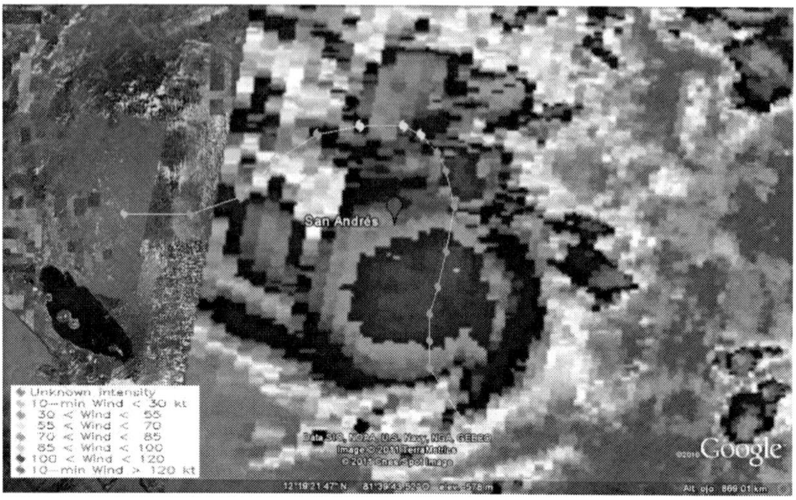

Figure 10. Historical image of Joan 1988 (October 19 00:00 Hrs), using Google Earth® and HURSAT (HURricane SATelite data).

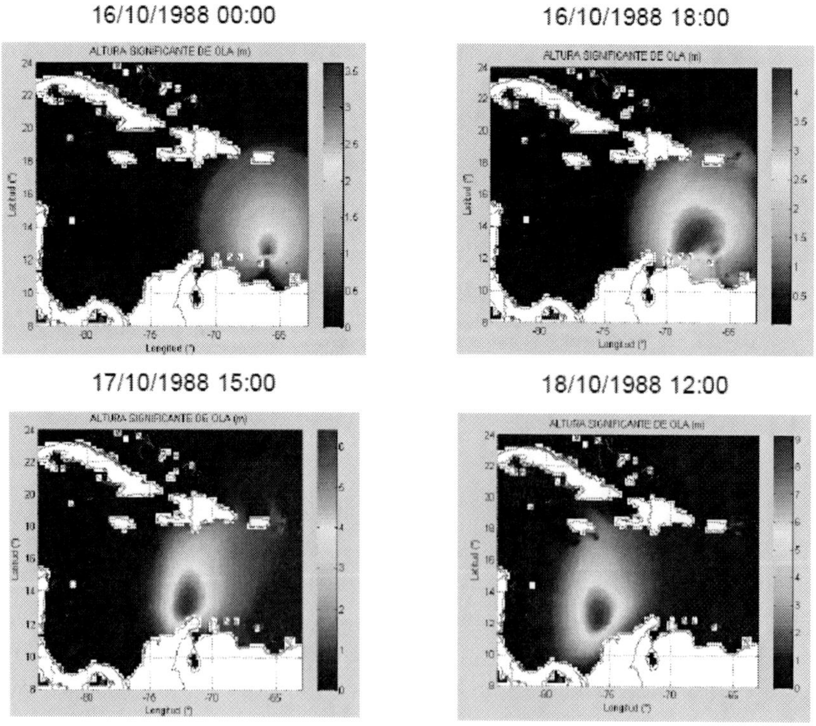

Figure 11. Waves simulated of Joan 1988 (October 19 00:00 Hrs).

Inundada la 'Heroica'

MARIA LUZ DUQUE

HASTA EL CENTRO AMURALLADO llegó el mar de leva que afecta a Cartagena, en donde inundó el baluarte de San Francisco Javier. El Joan continuaba su aumento, tanto así, que dirigió su rumbo a las costas de Nicaragua y Costa Rica.

Figure 12. Hurricane Joan 1988 (Image taken from *El Tiempo* news paper).

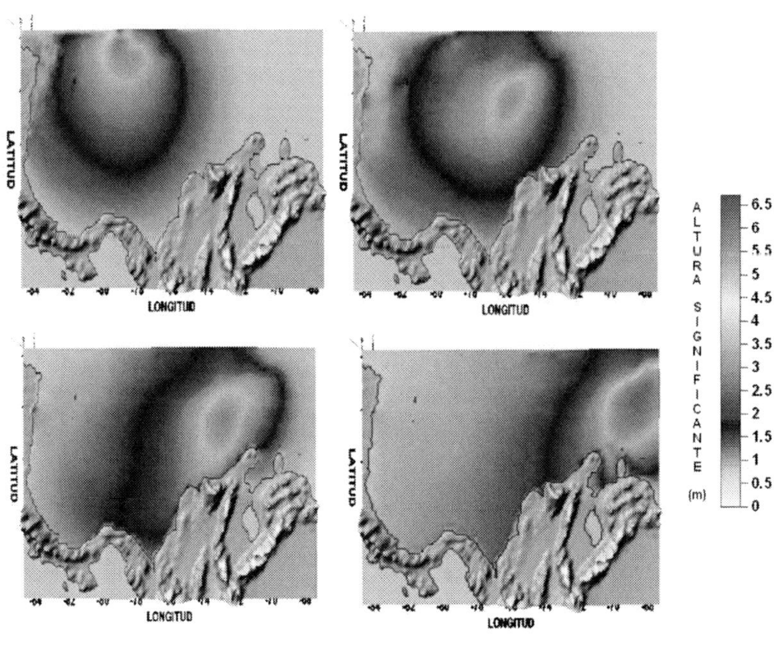

Figure 13. Waves simulation of hurricane Lenny in 1999 (Ortiz, 2009).

Tornadoes

In the other hand, the tornadoes and windstorms (called locally 'vendavales') have been reported in Barranquilla and the surrounding metropolitan area during the past 10 years. Likewise, the chapter involved monthly climate studies of the main meteorological variables at the airport station during the past 10 years. In the same manner, we conducted an air temperature analysis for the past 40 years. The data analysis reveals that tornados have occurred between May and September. These months correspond to the period characterized by the highest temperatures: average temperatures of 28.2°C and maximum temperatures of 33.2 °C. This is also the period characterized by the highest relative humidity and precipitation (84% and 50 mm, respectively). Although the highest figures for precipitation and humidity are found in in October, temperatures during this month are lower compared to May – September. Additionally, during the months in which tornados appear, surface winds are at their lowest level with respect to any other time of year. Because tornadoes have only appeared in the last 10 years (5 since 2001), is not possible to determine a definite relationship between their occurrence and large- scale climatic change Ortiz and Rosales (2012).

According to Ortiz and Rosales (2012), on June 1, 2001 the first tornado took place in the region. On 2010, four such events have occurred. On September 15, 2006, a tornado formed in Barranquilla causing severe damage to houses and buildings in a large part of the northeastern section of the city. This has been the most devastating event during the short history of tornadoes in Colombia. The Figure 14 shows the impact to the tornado in Barranquilla on 2006. Although weather phenomena characterized by sudden strong winds and gusts that cause damage to buildings (commonly referred to as gales or windstorms) have been common for decades in this part of Colombia, prior to 2001, no tornadoes had appeared in the region. From 2001 onward, the tornadoes have been forming continuously.

A total of 5 tornadoes have taken place in this region of Colombia. The dates of these tornadoes are as follows: June 1, 2001; September 15, 2006; June 5, 2007 and May 22, 2008. Prior to these dates, there has been no report of the appearance of tornadoes. During 2010, there were at least three phenomena that created vortices on the following dates: July 20, August 3 and August 12. None of them touched the ground. Recently, on June of 2012 the fifth tornado effected a zone near to Barranquilla, in the Sabanalarga town, located to 45 Km approximately from Barranquilla (Figure 15), one death and more of 100 houses were affected (Figure 16).

Figure 14 Tornado in Barranquilla, September 15, 2006. (Image taken from *El Heraldo*).

Figure 15. Distance from Sabanalarga and Barranquilla (Image taken from Google Earth® 2012).

Figure 16. Tornado in Sabanalarga, June 1, 2012. (Image taken from *El Heraldo).*

Cold Fronts

There is not so much information publicized about cold fronts in the Caribbean zone of Colombia. A recent study about of the collapse of Puerto Colombia pier due extreme waves produced by a cold front is a first work.

On March 7, 2009, a 200-meter section of the Puerto Colombia Pier collapsed during the passage of a cold front through the Atlántico Department of Colombia on the Caribbean coast (Figure 17).

In the study, (Ortiz et al., 2012), the most important meteo-marine event of the last ten years in the Colombian Caribbean area was reconstructed using SWAN and the Wavewatch III (WW3) global wave model, which uses wind fields as input. The modeling involved a nested grid approach for the generation and propagation of waves in a particular sector of the central Caribbean coast of Colombia. To validate the numerical model, the time series of the event generated using modeling data was compared with the recorded wave data of buoy 42058, located in the Caribbean Sea and operated by the National Oceanic and Atmospheric Administration (NOAA), and buoy 41194, located near the mouth of the Magdalena River and operated by the General Maritime Directorate (DIMAR).

Figure 17. Collapsed pier, March 07 of 2009, 07:00 local time, photo taken 2 hours after of the collapse (Ortiz et al 2012).

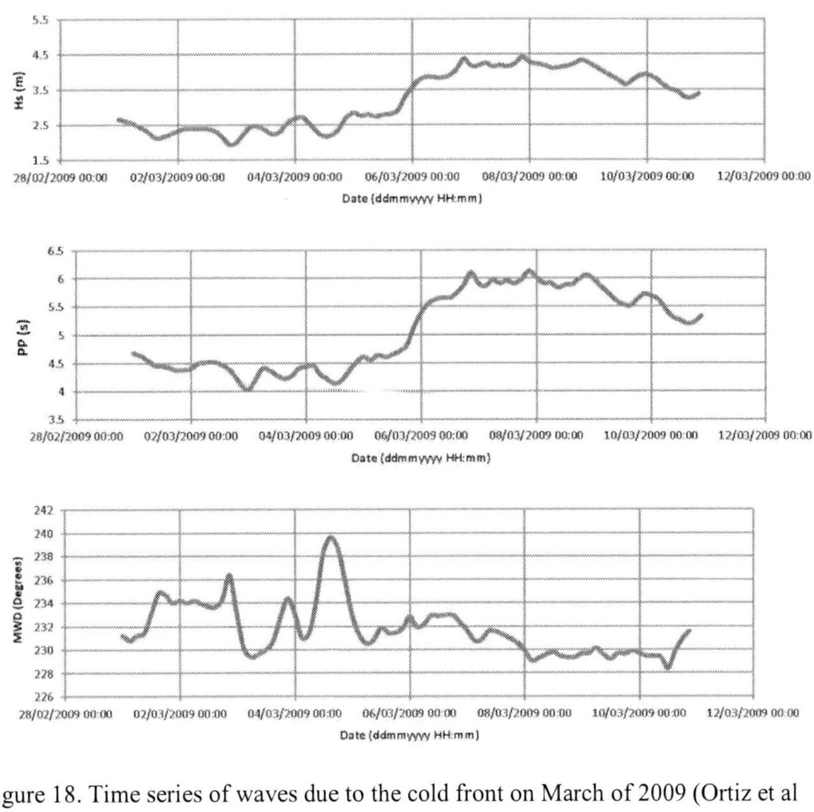

Figure 18. Time series of waves due to the cold front on March of 2009 (Ortiz et al 2012).

The simulation results were incorporated into the CMS-WAVE and CMS-FLOW models for the reconstruction of the cold-front event.

The modeling results were also compared to previously reported data for the waves generated by the hurricanes Joan (1988) and Lenny (1999). The results revealed that the analyzed study area was more heavily impacted by the ocean conditions produced by the 2009 cold front event than those produced by the hurricanes that have affected the Colombian Caribbean coast. A study conducted by Ortiz et al., (2008) modeled the wind fields and waves generated by Hurricane Joan in 1988, whose trajectory was in close proximity to the Colombian Caribbean coast. Although the total duration of the event (i.e., the time during which the tropical depression developed into a hurricane then dissipated) was thirteen days, only two of those days had any impact on the study area. The modeling indicated the occurrence of waves with heights up to 2.4 m in the sector of interest (Figure 18).

Acknowledgments

The authors would like to thank the Office of Research and Projects (DIDI) at the Universidad del Norte for the support, and the Dr. Mauro Maza for your suggestions.

References

Alvarado M (2009). *Río Magdalena navegación marítima y fluvial* (1986-2008). Barranquilla, Ediciones Uninorte.

Andrade, C.A. (2001). Las corrientes superficiales en la Cuenca de Colombia observadas con boyas de deriva. *Rev. Acad. Colomb. Cienc* 25 (96):321-335.

Andrade C. and Barton E (2001). Eddy development and motion in the Caribbean Sea. *J. Geo. Res*: 105(C11):26191-26201

Bernal G., Poveda G., Roldán P. and Andrade C (2006). Patrones de variabilidad de las temperaturas superficiales del mar en la Costa Caribe Colombiana. *Rev. Acad. Colomb. Cienc*. 30 (115): 195-208.

El Tiempo News Paper [website] [Updated 2011 November 9; Cited 2011 November 9]. Available at: *http://www.eltiempo.com/seccion_archivo/ index.php*.

El Heraldo. July 21, 2011, Barranquilla, Colombia.

El Heraldo. September 17, 2006, Barranquilla, Colombia.

Invemar (2010). Biodiversidad del margen continental del Caribe colombiano. *Invemar* 20: 54.

Meisel A (2003). The Continentalization of the Island of San Andrés, Colombia: Panyas, raizales and tourism, 1953-2003. *Working Documents on Local, N° 37, Bank of the Republic*, CEER-Cartagena.

Ortiz J.C. and Rosales M (2012). Severe Tornadoes on the Caribbean Coast of Colombia Since 2001 and Their Relation to Local Climate Conditions. *Natural Hazard Journal* DOI: 0.1007/s11069-012-0337-8.

Ortiz J.C., Salcedo B. and Otero L (2012). Investigating the collapse of the Puerto Colombia Pier (Colombian Caribbean Coast) in March of 2009: methodology for the reconstruction of extreme events and their impact on the coastal infrastructure. *Journal of Coastal Research* (on press).

Ortiz J.C. (2012). Exposure of the Colombian Caribbean Coast, including San Andrés Island, to Tropical Storms and Hurricanes, 1900–2010. *Natural Hazard Journal* (2012) 61:815-827.

Ortiz J.C (2007). Hurricanes and Storms in the Colombian Caribbean since 1900. *Boletín Científico, CIOH,* No. 25: 54-60.

Ortiz, .J.C., López Martinez, F., Diaz Estrada, E. and Bacca Arango, L.F (2008). Estudio de Oleaje Generado por el Huracán Joan en la Costa Caribe Colombiana en 1988, Incluyendo la Isla de San Andres. *Revista Colombiana de Física*, 40(2), pp.444-46.

Ortiz, J.C (2009). Aplicación de un modelo paramétrico de vientos y un modelo de oleaje espectral en el estudio del oleaje máximo generado por el huracan Lenny, en las costas del caribe Colombiano. *Boletín Científico CIOH*, 27, pp.29-36.

Osorio, A.F., Mesa, J.C. and Bernal, G (2009). Reconstrucción de cuarenta años de datos de oleaje en el Mar Caribe Colombiano empleando el Modelo WWIII y diferentes fuentes de datos. *Boletín Científico CIOH*, 27, pp.37-56.

Viloria J (2006). Port Cities of the Colombian Caribbean: Proposals for Competition in a Globalized Economy. *Working Documents on Regional Economics, N° 80, Bank of the Republic*, CEER-Cartagena.

INDEX

C

D

E

I

J

K

L

M

N

O

P

Q

R

S

T

U

V

W

Y

Z